A Member of the International Code Family™

INTERNATIONAL MECHANICAL CODE®

INTERNATIONAL
CODE COUNCIL®

2003

2003 International Mechanical Code®

First Printing: February 2003
Second Printing: May 2003

ISBN # 1-892395-64-9 (soft-cover edition)
ISBN # 1-892395-63-0 (loose-leaf edition)
ISBN # 1-892395-83-5 (e-document)

PRINTED IN THE U.S.A.

PREFACE

Introduction

Internationally, code officials recognize the need for a modern, up-to-date mechanical code addressing the design and installation of mechanical systems emphasizing performance. The *International Mechanical Code*®, in this 2003 edition, is designed to meet these needs through model code regulations that safeguard the public health and safety in all communities, large and small.

This comprehensive mechanical code establishes minimum regulations for mechanical systems using prescriptive and performance-related provisions. It is founded on broad-based principles that make possible the use of new materials and new mechanical system designs. This 2003 edition is fully compatible with all the *International Codes* ("I-Codes") published by the International Code Council (ICC), including the *International Building Code, ICC Electrical Code, International Energy Conservation Code, International Existing Building Code, International Fire Code, International Fuel Gas Code, ICC Performance Code, International Plumbing Code, International Private Sewage Disposal Code, International Property Maintenance Code, International Residential Code, International Urban-Wildland Interface Code* and *International Zoning Code.*

The *International Mechanical Code* provisions provide many benefits, among which is the model code development process that offers an international forum for mechanical professionals to discuss performance and prescriptive code requirements. This forum provides an excellent arena to debate proposed revisions. This model code also encourages international consistency in the application of provisions.

Development

The first edition of the *International Mechanical Code* (1996) was the culmination of an effort initiated in 1995 by a development committee appointed by the International Code Council (ICC) and consisting of representatives of the three statutory members of the ICC: Building Officials and Code Administrators International, Inc. (BOCA), International Conference of Building Officials (ICBO) and Southern Building Code Congress International (SBCCI). The intent was to draft a comprehensive set of regulations for mechanical systems consistent with and inclusive of the scope of the existing model codes. Technical content of the latest model codes promulgated by BOCA, ICBO and SBCCI was utilized as the basis for the development. This 2003 edition presents the code as originally issued, with changes approved through the ICC Code Development Process through 2002. A new edition such as this is promulgated every three years.

With the development and publication of the family of *International Codes* in 2000, the continued development and maintenance of the model codes individually promulgated by BOCA ("BOCA National Codes"), ICBO ("Uniform Codes") and SBCCI ("Standard Codes") was discontinued. This 2003 *International Mechanical Code*, as well as its predecessor–the 2000 edition–is intended to be the successor mechanical code to those codes previously developed by BOCA, ICBO and SBCCI.

The development of a single set of comprehensive and coordinated *International Codes* was a significant milestone in the development of regulations for the built environment. The timing of this publication mirrors a milestone in the change in structure of the model codes, namely, the consolidation of BOCA, ICBO and SBCCI into the ICC. The activities and services previously provided by the individual model code organizations will be the responsibility of the consolidated ICC.

This code is founded on principles intended to establish provisions consistent with the scope of a mechanical code that adequately protects public health, safety and welfare; provisions that do not unnecessarily increase construction costs; provisions that do not restrict the use of new materials, products or methods of construction; and provisions that do not give preferential treatment to particular types or classes of materials, products or methods of construction.

Adoption

The *International Mechanical Code* is available for adoption and use by jurisdictions internationally. Its use within a governmental jurisdiction is intended to be accomplished through adoption by reference in accordance with proceedings establishing the jurisdiction's laws. At the time of adoption, jurisdictions should insert the appropriate information in provisions requiring specific local information, such as the name of the adopting jurisdiction. These locations are shown in bracketed words in small capital letters in the code and in the sample ordinance. The sample adoption ordinance on page v addresses several key elements of a code adoption ordinance, including the information required for insertion into the code text.

Maintenance

The *International Mechanical Code* is kept up to date through the review of proposed changes submitted by code enforcing officials, industry representatives, design professionals and other interested parties. Proposed changes are carefully considered through an open code development process in which all interested and affected parties may participate.

The contents of this work are subject to change both through the Code Development Cycles and the governmental body that enacts the code into law. For more information regarding the code development process, contact the Code and Standard Development Department of the International Code Council.

While the development procedure of the *International Mechanical Code* assures the highest degree of care, ICC and the founding members of ICC—BOCA, ICBO, SBCCI—their members and those participating in the development of this code do not accept any liability resulting from compliance or noncompliance with the provisions because ICC and its founding members do not have the power or authority to police or enforce compliance with the contents of this code. Only the governmental body that enacts the code into law has such authority.

Letter Designations in Front of Section Numbers

In each code development cycle, proposed changes to this code are considered at the Code Development Hearing by the International Mechanical Code Development Committee, whose action constitutes a recommendation to the voting membership for final action on the proposed change. Proposed changes to a code section whose number begins with a letter in brackets are considered by a different code development committee. For instance, proposed changes to code sections which have the letter [B] in front (for example, [B] 601.2), are considered by the International Building Code Development Committee at the Code Development Hearing. Where this designation is applicable to the entire content of a main section of the code, the designation appears at the main section number and title and is not repeated at every subsection in that section.

The content of sections in this code which begin with a letter designation are maintained by another code development committee in accordance with the following: [B]= International Building Code Development Committee; [E] = International Energy Conservation Code Development Committee; [EB] = International Existing Building Code Development Committee; and [F] = International Fire Code Development Committee.

Marginal Markings

Solid vertical lines in the margins within the body of the code indicate a technical change from the requirements of the 2000 edition. Deletion indicators (➤) are provided in the margin where a paragraph or item has been deleted.

ORDINANCE

The *International Codes* are designed and promulgated to be adopted by reference by ordinance. Jurisdictions wishing to adopt the 2003 *International Mechanical Code* as an enforceable regulation governing mechanical systems should ensure that certain factual information is included in the adopting ordinance at the time adoption is being considered by the appropriate governmental body. The following sample adoption ordinance addresses several key elements of a code adoption ordinance, including the information required for insertion into the code text.

SAMPLE ORDINANCE FOR ADOPTION OF
THE *INTERNATIONAL MECHANICAL CODE*
ORDINANCE NO._____

An ordinance of the [JURISDICTION] adopting the 2003 edition of the *International Mechanical Code*, regulating and governing the design, construction, quality of materials, erection, installation, alteration, repair, location, relocation, replacement, addition to, use or maintenance of mechanical systems in the [JURISDICTION]; providing for the issuance of permits and collection of fees therefor; repealing Ordinance No. _____ of the [JURISDICTION] and all other ordinances and parts of the ordinances in conflict therewith.

The [GOVERNING BODY] of the [JURISDICTION] does ordain as follows:

Section 1.That a certain document, three (3) copies of which are on file in the office of the [TITLE OF JURISDICTION'S KEEPER OF RECORDS] OF [NAME OF JURISDICTION] , being marked and designated as the *International Mechanical Code*, 2003 edition, including Appendix Chapters [FILL IN THE APPENDIX CHAPTERS BEING ADOPTED] (see *International Mechanical Code* Section 101.2.1, 2003 edition), as published by the International Code Council, be and is hereby adopted as the Mechanical Code of the [JURISDICTION], in the State of [STATE NAME] regulating and governing the design, construction, quality of materials, erection, installation, alteration, repair, location, relocation, replacement, addition to, use or maintenance of mechanical systems as herein provided; providing for the issuance of permits and collection of fees therefor; and each and all of the regulations, provisions, penalties, conditions and terms of said Mechanical Code on file in the office of the [JURISDICTION] are hereby referred to, adopted, and made a part hereof, as if fully set out in this ordinance, with the additions, insertions, deletions and changes, if any, prescribed in Section 2 of this ordinance.

Section 2.The following sections are hereby revised:

Section 101.1. Insert: [NAME OF JURISDICTION]

Section 106.5.2. Insert: [APPROPRIATE SCHEDULE]

Section 106.5.3. Insert: [PERCENTAGES IN TWO LOCATIONS]

Section 108.4. Insert: [OFFENSE, DOLLAR AMOUNT, NUMBER OF DAYS]

Section 108.5. Insert: [DOLLAR AMOUNT IN TWO LOCATIONS]

Section 3.That Ordinance No. _____ of [JURISDICTION] entitled [FILL IN HERE THE COMPLETE TITLE OF THE ORDINANCE OR ORDINANCES IN EFFECT AT THE PRESENT TIME SO THAT THEY WILL BE REPEALED BY DEFINITE MENTION] and all other ordinances or parts of ordinances in conflict herewith are hereby repealed.

Section 4.That if any section, subsection, sentence, clause or phrase of this ordinance is, for any reason, held to be unconstitutional, such decision shall not affect the validity of the remaining portions of this ordinance. The [GOVERNING BODY] hereby declares that it would have passed this ordinance, and each section, subsection, clause or phrase thereof, irrespective of the fact that any one or more sections, subsections, sentences, clauses and phrases be declared unconstitutional.

Section 5.That nothing in this ordinance or in the Mechanical Code hereby adopted shall be construed to affect any suit or proceeding impending in any court, or any rights acquired, or liability incurred, or any cause or causes of action acquired or existing, under any act or ordinance hereby repealed as cited in Section 2 of this ordinance; nor shall any just or legal right or remedy of any character be lost, impaired or affected by this ordinance.

Section 6.That the [JURISDICTION'S KEEPER OF RECORDS] is hereby ordered and directed to cause this ordinance to be published. (An additional provision may be required to direct the number of times the ordinance is to be published and to specify that it is to be in a newspaper in general circulation. Posting may also be required.)

Section 7.That this ordinance and the rules, regulations, provisions, requirements, orders and matters established and adopted hereby shall take effect and be in full force and effect [TIME PERIOD] from and after the date of its final passage and adoption.

TABLE OF CONTENTS

CHAPTER 1

ADMINISTRATION

SECTION 101
GENERAL

101.1 Title. These regulations shall be known as the *Mechanical Code* of [NAME OF JURISDICTION], hereinafter referred to as "this code."

101.2 Scope. This code shall regulate the design, installation, maintenance, alteration and inspection of mechanical systems that are permanently installed and utilized to provide control of environmental conditions and related processes within buildings. This code shall also regulate those mechanical systems, system components, equipment and appliances specifically addressed herein. The installation of fuel gas distribution piping and equipment, fuel gas-fired appliances and fuel gas-fired appliance venting systems shall be regulated by the *International Fuel Gas Code.*

Exceptions:

1. Detached one- and two-family dwellings and multiple single-family dwellings (townhouses) not more than three stories high with separate means of egress and their accessory structures shall comply with the *International Residential Code.*

2. Mechanical systems in existing buildings undergoing repair, alterations, or additions, and change of occupancy shall be permitted to comply with the *International Existing Building Code.*

101.2.1 Appendices. Provisions in the appendices shall not apply unless specifically adopted.

101.3 Intent. The purpose of this code is to provide minimum standards to safeguard life or limb, health, property and public welfare by regulating and controlling the design, construction, installation, quality of materials, location, operation and maintenance or use of mechanical systems.

101.4 Severability. If a section, subsection, sentence, clause or phrase of this code is, for any reason, held to be unconstitutional, such decision shall not affect the validity of the remaining portions of this code.

SECTION 102
APPLICABILITY

102.1 General. The provisions of this code shall apply to all matters affecting or relating to structures and premises, as set forth in Section 101. Where, in a specific case, different sections of this code specify different materials, methods of construction or other requirements, the most restrictive shall govern.

102.2 Existing installations. Except as otherwise provided for in this chapter, a provision in this code shall not require the removal, alteration or abandonment of, nor prevent the continued utilization and maintenance of, a mechanical system lawfully in existence at the time of the adoption of this code.

102.3 Maintenance. Mechanical systems, both existing and new, and parts thereof shall be maintained in proper operating condition in accordance with the original design and in a safe and sanitary condition. Devices or safeguards which are required by this code shall be maintained in compliance with the code edition under which they were installed. The owner or the owner's designated agent shall be responsible for maintenance of mechanical systems. To determine compliance with this provision, the code official shall have the authority to require a mechanical system to be reinspected.

[EB] 102.4 Additions, alterations or repairs. Additions, alterations, renovations or repairs to a mechanical system shall conform to that required for a new mechanical system without requiring the existing mechanical system to comply with all of the requirements of this code. Additions, alterations or repairs shall not cause an existing mechanical system to become unsafe, hazardous or overloaded.

Minor additions, alterations, renovations and repairs to existing mechanical systems shall meet the provisions for new construction, unless such work is done in the same manner and arrangement as was in the existing system, is not hazardous and is approved.

[EB] 102.5 Change in occupancy. It shall be unlawful to make a change in the occupancy of any structure which will subject the structure to any special provision of this code applicable to the new occupancy without approval. The code official shall certify that such structure meets the intent of the provisions of law governing building construction for the proposed new occupancy and that such change of occupancy does not result in any hazard to the public health, safety or welfare.

[EB] 102.6 Historic buildings. The provisions of this code relating to the construction, alteration, repair, enlargement, restoration, relocation or moving of buildings or structures shall not be mandatory for existing buildings or structures identified and classified by the state or local jurisdiction as historic buildings when such buildings or structures are judged by the code official to be safe and in the public interest of health, safety and welfare regarding any proposed construction, alteration, repair, enlargement, restoration, relocation or moving of buildings.

102.7 Moved buildings. Except as determined by Section 102.2, mechanical systems that are a part of buildings or structures moved into or within the jurisdiction shall comply with the provisions of this code for new installations.

102.8 Referenced codes and standards. The codes and standards referenced herein shall be those that are listed in Chapter 15 and such codes and standards shall be considered as part of the requirements of this code to the prescribed extent of each such reference. Where differences occur between provisions of this code and the referenced standards, the provisions of this code shall apply.

102.9 Requirements not covered by this code. Requirements necessary for the strength, stability or proper operation of an existing or proposed mechanical system, or for the public safety, health and general welfare, not specifically covered by this code, shall be determined by the code official.

SECTION 103
DEPARTMENT OF MECHANICAL INSPECTION

103.1 General. The department of mechanical inspection is hereby created and the executive official in charge thereof shall be known as the code official.

103.2 Appointment. The code official shall be appointed by the chief appointing authority of the jurisdiction; and the code official shall not be removed from office except for cause and after full opportunity to be heard on specific and relevant charges by and before the appointing authority.

103.3 Deputies. In accordance with the prescribed procedures of this jurisdiction and with the concurrence of the appointing authority, the code official shall have the authority to appoint a deputy code official, other related technical officers, inspectors and other employees.

103.4 Liability. The code official, officer or employee charged with the enforcement of this code, while acting for the jurisdiction, shall not thereby be rendered liable personally, and is hereby relieved from all personal liability for any damage accruing to persons or property as a result of an act required or permitted in the discharge of official duties.

Any suit instituted against any officer or employee because of an act performed by that officer or employee in the lawful discharge of duties and under the provisions of this code shall be defended by the legal representative of the jurisdiction until the final termination of the proceedings. The code official or any subordinate shall not be liable for costs in an action, suit or proceeding that is instituted in pursuance of the provisions of this code; and any officer of the department of mechanical inspection, acting in good faith and without malice, shall be free from liability for acts performed under any of its provisions or by reason of any act or omission in the performance of official duties in connection therewith.

SECTION 104
DUTIES AND POWERS OF THE CODE OFFICIAL

104.1 General. The code official shall enforce the provisions of this code and shall act on any question relative to the installation, alteration, repair, maintenance or operation of mechanical systems, except as otherwise specifically provided for by statutory requirements or as provided for in Sections 104.2 through 104.8.

104.2 Rule-making authority. The code official shall have authority as necessary in the interest of public health, safety and general welfare, to adopt and promulgate rules and regulations; to interpret and implement the provisions of this code; to secure the intent thereof; and to designate requirements applicable because of local climatic or other conditions. Such rules shall not have the effect of waiving structural or fire performance re-quirements specifically provided for in this code, or of violating accepted engineering methods involving public safety.

104.3 Applications and permits. The code official shall receive applications and issue permits for the installation and alteration of mechanical systems, inspect the premises for which such permits have been issued and enforce compliance with the provisions of this code.

104.4 Inspections. The code official shall make all of the required inspections, or shall accept reports of inspection by approved agencies or individuals. All reports of such inspections shall be in writing and be certified by a responsible officer of such approved agency or by the responsible individual. The code official is authorized to engage such expert opinion as deemed necessary to report upon unusual technical issues that arise, subject to the approval of the appointing authority.

104.5 Right of entry. Whenever it is necessary to make an inspection to enforce the provisions of this code, or whenever the code official has reasonable cause to believe that there exists in a building or upon any premises any conditions or violations of this code which make the building or premises unsafe, insanitary, dangerous or hazardous, the code official shall have the authority to enter the building or premises at all reasonable times to inspect or to perform the duties imposed upon the code official by this code. If such building or premises is occupied, the code official shall present credentials to the occupant and request entry. If such building or premises is unoccupied, the code official shall first make a reasonable effort to locate the owner or other person having charge or control of the building or premises and request entry. If entry is refused, the code official has recourse to every remedy provided by law to secure entry.

When the code official has first obtained a proper inspection warrant or other remedy provided by law to secure entry, an owner or occupant or person having charge, care or control of the building or premises shall not fail or neglect, after proper request is made as herein provided, to promptly permit entry therein by the code official for the purpose of inspection and examination pursuant to this code.

104.6 Identification. The code official shall carry proper identification when inspecting structures or premises in the performance of duties under this code.

104.7 Notices and orders. The code official shall issue all necessary notices or orders to ensure compliance with this code.

104.8 Department records. The code official shall keep official records of applications received, permits and certificates issued, fees collected, reports of inspections, and notices and orders issued. Such records shall be retained in the official records as long as the building or structure to which such records relate remains in existence, unless otherwise provided for by other regulations.

SECTION 105
APPROVAL

105.1 Modifications. Whenever there are practical difficulties involved in carrying out the provisions of this code, the code official shall have the authority to grant modifications for individual cases, provided the code official shall first find that special individual reason makes the strict letter of this code impractical and the modification is in compliance with the intent and purpose of this code and does not lessen health, life and fire safety requirements. The details of action granting modifications shall be recorded and entered in the files of the mechanical inspection department.

105.2 Alternative materials, methods, equipment and appliances. The provisions of this code are not intended to prevent the installation of any material or to prohibit any method of construction not specifically prescribed by this code, provided that any such alternative has been approved. An alternative material or method of construction shall be approved where the code official finds that the proposed design is satisfactory and complies with the intent of the provisions of this code, and that the material, method or work offered is, for the purpose intended, at least the equivalent of that prescribed in this code in quality, strength, effectiveness, fire resistance, durability and safety.

105.3 Required testing. Whenever there is insufficient evidence of compliance with the provisions of this code, or evidence that a material or method does not conform to the requirements of this code, or in order to substantiate claims for alternative materials or methods, the code official shall have the authority to require tests as evidence of compliance to be made at no expense to the jurisdiction.

105.3.1 Test methods. Test methods shall be as specified in this code or by other recognized test standards. In the absence of recognized and accepted test methods, the code official shall approve the testing procedures.

105.3.2 Testing agency. All tests shall be performed by an approved agency.

105.3.3 Test reports. Reports of tests shall be retained by the code official for the period required for retention of public records.

105.4 Material, equipment and appliance reuse. Materials, equipment, appliances and devices shall not be reused unless such elements have been reconditioned, tested and placed in good and proper working condition and approved.

SECTION 106
PERMITS

106.1 When required. An owner, authorized agent or contractor who desires to erect, install, enlarge, alter, repair, remove, convert or replace a mechanical system, the installation of which is regulated by this code, or to cause such work to be done, shall first make application to the code official and obtain the required permit for the work.

Exception: Where equipment and appliance replacements or repairs must be performed in an emergency situation, the permit application shall be submitted within the next working business day of the department of mechanical inspection.

106.2 Permits not required. Permits shall not be required for the following:

1. Portable heating appliances;

2. Portable ventilation appliances and equipment;

3. Portable cooling units;

4. Steam, hot water or chilled water piping within any heating or cooling equipment or appliances regulated by this code;

5. The replacement of any minor part that does not alter the approval of equipment or an appliance or make such equipment or appliance unsafe;

6. Portable evaporative coolers; and

7. Self-contained refrigeration systems that contain 10 pounds (4.5 kg) or less of refrigerant, or that are actuated by motors of 1 horsepower (0.75 kW) or less.

8. Portable fuel cell appliances that are not connected to a fixed piping system and are not interconnected to a power grid.

Exemption from the permit requirements of this code shall not be deemed to grant authorization for work to be done in violation of the provisions of this code or other laws or ordinances of this jurisdiction.

106.3 Application for permit. Each application for a permit, with the required fee, shall be filed with the code official on a form furnished for that purpose and shall contain a general description of the proposed work and its location. The application shall be signed by the owner or an authorized agent. The permit application shall indicate the proposed occupancy of all parts of the building and of that portion of the site or lot, if any, not covered by the building or structure and shall contain such other information required by the code official.

106.3.1 Construction documents. Construction documents, engineering calculations, diagrams and other data shall be submitted in two or more sets with each application for a permit. The code official shall require construction documents, computations and specifications to be prepared and designed by a registered design professional when required by state law. Construction documents shall be drawn to scale and shall be of sufficient clarity to indicate the location, nature and extent of the work proposed and show in detail that the work conforms to the provisions of this code. Construction documents for buildings more than two stories in height shall indicate where penetrations will be made for mechanical systems, and the materials and methods for maintaining required structural safety, fire-resistance rating and fireblocking.

Exception: The code official shall have the authority to waive the submission of construction documents, calculations or other data if the nature of the work applied for is such that reviewing of construction documents is not necessary to determine compliance with this code.

106.4 Permit issuance. The application, construction documents and other data filed by an applicant for a permit shall be reviewed by the code official. If the code official finds that the proposed work conforms to the requirements of this code and all laws and ordinances applicable thereto, and that the fees specified in Section 106.5 have been paid, a permit shall be issued to the applicant.

106.4.1 Approved construction documents. When the code official issues the permit where construction documents are required, the construction documents shall be endorsed in writing and stamped "APPROVED." Such approved construction documents shall not be changed, modified or altered without authorization from the code of-ficial. Work shall be done in accordance with the approved construction documents.

The code official shall have the authority to issue a permit for the construction of part of a mechanical system before the construction documents for the entire system have been submitted or approved, provided adequate information and detailed statements have been filed complying with all pertinent requirements of this code. The holder of such permit shall proceed at his or her own risk without assurance that the permit for the entire mechanical system will be granted.

106.4.2 Validity. The issuance of a permit or approval of construction documents shall not be construed to be a permit for, or an approval of, any violation of any of the provisions of this code or of other ordinances of the jurisdiction. A permit presuming to give authority to violate or cancel the provisions of this code shall be invalid.

The issuance of a permit based upon construction documents and other data shall not prevent the code official from thereafter requiring the correction of errors in said construction documents and other data or from preventing building operations from being carried on thereunder when in violation of this code or of other ordinances of this jurisdiction.

106.4.3 Expiration. Every permit issued by the code official under the provisions of this code shall expire by limitation and become null and void if the work authorized by such permit is not commenced within 180 days from the date of such permit, or if the work authorized by such permit is suspended or abandoned at any time after the work is commenced for a period of 180 days. Before such work recommences, a new permit shall be first obtained and the fee, therefore, shall be one-half the amount required for a new permit for such work, provided no changes have been made or will be made in the original construction documents for such work, and provided further that such suspension or abandonment has not exceeded one year.

106.4.4 Extensions. A permittee holding an unexpired permit shall have the right to apply for an extension of the time within which the permittee will commence work under that permit when work is unable to be commenced within the time required by this section for good and satisfactory reasons. The code official shall extend the time for action by the permittee for a period not exceeding 180 days if there is reasonable cause. A permit shall not be extended more than once. The fee for an extension shall be one-half the amount required for a new permit for such work.

106.4.5 Suspension or revocation of permit. The code official shall revoke a permit or approval issued under the provisions of this code in case of any false statement or misrepresentation of fact in the application or on the construction documents upon which the permit or approval was based.

106.4.6 Retention of construction documents. One set of construction documents shall be retained by the code official until final approval of the work covered therein. One set of approved construction documents shall be returned to the

applicant, and said set shall be kept on the site of the building or job at all times during which the work authorized thereby is in progress.

106.5 Fees. A permit shall not be issued until the fees prescribed in Section 106.5.2 have been paid, nor shall an amendment to a permit be released until the additional fee, if any, due to an increase of the mechanical system, has been paid.

106.5.1 Work commencing before permit issuance. Any person who commences work on a mechanical system before obtaining the necessary permits shall be subject to 100 percent of the usual permit fee in addition to the required permit fees.

106.5.2 Fee schedule. The fees for mechanical work shall be as indicated in the following schedule.

[JURISDICTION TO INSERT
APPROPRIATE SCHEDULE]

106.5.3 Fee refunds. The code official shall authorize the refunding of fees as follows.

1. The full amount of any fee paid hereunder which was erroneously paid or collected.

2. Not more than [SPECIFY PERCENTAGE] percent of the permit fee paid when no work has been done under a permit issued in accordance with this code.

3. Not more than [SPECIFY PERCENTAGE] percent of the plan review fee paid when an application for a permit for which a plan review fee has been paid is withdrawn or canceled before any plan review effort has been expended.

The code official shall not authorize the refunding of any fee paid, except upon written application filed by the original permittee not later than 180 days after the date of fee payment.

SECTION 107
INSPECTIONS AND TESTING

107.1 Required inspections and testing. The code official, upon notification from the permit holder or the permit holder's agent, shall make the following inspections and other such inspections as necessary, and shall either release that portion of the construction or shall notify the permit holder or the permit holder's agent of violations that must be corrected. The holder of the permit shall be responsible for the scheduling of such inspections.

1. Underground inspection shall be made after trenches or ditches are excavated and bedded, piping installed, and before backfill is put in place. When excavated soil contains rocks, broken concrete, frozen chunks and other rubble that would damage or break the piping or cause corrosive action, clean backfill shall be on the job site.

2. Rough-in inspection shall be made after the roof, framing, fireblocking and bracing are in place and all ducting and other components to be concealed are complete, and prior to the installation of wall or ceiling membranes.

3. Final inspection shall be made upon completion of the mechanical system.

Exception: Ground-source heat pump loop systems tested in accordance with Section 1208.1.1 shall be permitted to be backfilled prior to inspection.

The requirements of this section shall not be considered to prohibit the operation of any heating equipment or appliances installed to replace existing heating equipment or appliances serving an occupied portion of a structure provided that a request for inspection of such heating equipment or appliances has been filed with the department not more than 48 hours after such replacement work is completed, and before any portion of such equipment or appliances is concealed by any permanent portion of the structure.

107.1.1 Approved inspection agencies. The code official shall accept reports of approved agencies, provided that such agencies satisfy the requirements as to qualifications and reliability.

107.1.2 Evaluation and follow-up inspection services. Prior to the approval of a prefabricated construction assembly having concealed mechanical work and the issuance of a mechanical permit, the code official shall require the submittal of an evaluation report on each prefabricated construction assembly, indicating the complete details of the mechanical system, including a description of the system and its components, the basis upon which the system is being evaluated, test results and similar information, and other data as necessary for the code official to determine conformance to this code.

107.1.2.1 Evaluation service. The code official shall designate the evaluation service of an approved agency as the evaluation agency, and review such agency's evaluation report for adequacy and conformance to this code.

107.1.2.2 Follow-up inspection. Except where ready access is provided to mechanical systems, service equipment and accessories for complete inspection at the site without disassembly or dismantling, the code official shall conduct the in-plant inspections as frequently as necessary to ensure conformance to the approved evaluation report or shall designate an independent, approved inspection agency to conduct such inspections. The inspection agency shall furnish the code official with the follow-up inspection manual and a report of inspections upon request, and the mechanical system shall have an identifying label permanently affixed to the system indicating that factory inspections have been performed.

107.1.2.3 Test and inspection records. Required test and inspection records shall be available to the code official at all times during the fabrication of the mechanical system and the erection of the building; or such records as the code official designates shall be filed.

107.2 Testing. Mechanical systems shall be tested as required in this code and in accordance with Sections 107.2.1 through 107.2.3. Tests shall be made by the permit holder and observed by the code official.

107.2.1 New, altered, extended or repaired systems. New mechanical systems and parts of existing systems, which have been altered, extended, renovated or repaired, shall be tested as prescribed herein to disclose leaks and defects.

107.2.2 Apparatus, material and labor for tests. Apparatus, material and labor required for testing a mechanical system or part thereof shall be furnished by the permit holder.

107.2.3 Reinspection and testing. Where any work or installation does not pass an initial test or inspection, the necessary corrections shall be made so as to achieve compliance with this code. The work or installation shall then be resubmitted to the code official for inspection and testing.

107.3 Approval. After the prescribed tests and inspections indicate that the work complies in all respects with this code, a notice of approval shall be issued by the code official.

107.4 Temporary connection. The code official shall have the authority to authorize the temporary connection of a mechanical system to the sources of energy for the purpose of testing mechanical systems or for use under a temporary certificate of occupancy.

SECTION 108
VIOLATIONS

108.1 Unlawful acts. It shall be unlawful for a person, firm or corporation to erect, construct, alter, repair, remove, demolish or utilize a mechanical system, or cause same to be done, in conflict with or in violation of any of the provisions of this code.

108.2 Notice of violation. The code official shall serve a notice of violation or order to the person responsible for the erection, installation, alteration, extension, repair, removal or demolition of mechanical work in violation of the provisions of this code, or in violation of a detail statement or the approved construction documents thereunder, or in violation of a permit or certificate issued under the provisions of this code. Such order shall direct the discontinuance of the illegal action or condition and the abatement of the violation.

108.3 Prosecution of violation. If the notice of violation is not complied with promptly, the code official shall request the legal counsel of the jurisdiction to institute the appropriate proceeding at law or in equity to restrain, correct or abate such violation, or to require the removal or termination of the unlawful occupancy of the structure in violation of the provisions of this code or of the order or direction made pursuant thereto.

108.4 Violation penalties. Persons who shall violate a provision of this code or shall fail to comply with any of the requirements thereof or who shall erect, install, alter or repair mechanical work in violation of the approved construction documents or directive of the code official, or of a permit or certificate issued under the provisions of this code, shall be guilty of a [SPECIFY OFFENSE], punishable by a fine of not more than [AMOUNT] dollars or by imprisonment not exceeding [NUMBER OF DAYS], or both such fine and imprisonment. Each day that a violation continues after due notice has been served shall be deemed a separate offense.

108.5 Stop work orders. Upon notice from the code official that mechanical work is being done contrary to the provisions of this code or in a dangerous or unsafe manner, such work shall immediately cease. Such notice shall be in writing and shall be given to the owner of the property, or to the owner's agent, or to the person doing the work. The notice shall state the conditions under which work is authorized to resume. Where an emergency exists, the code official shall not be required to give a written notice prior to stopping the work. Any person who shall continue any work on the system after having been served with a stop work order, except such work as that person is directed to perform to remove a violation or unsafe condition, shall be liable for a fine of not less than [AMOUNT] dollars or more than [AMOUNT] dollars.

108.6 Abatement of violation. The imposition of the penalties herein prescribed shall not preclude the legal officer of the jurisdiction from instituting appropriate action to prevent unlawful construction or to restrain, correct or abate a violation, or to prevent illegal occupancy of a building, structure or premises, or to stop an illegal act, conduct, business or utilization of the mechanical system on or about any premises.

108.7 Unsafe mechanical systems. A mechanical system that is unsafe, constitutes a fire or health hazard, or is otherwise dangerous to human life, as regulated by this code, is hereby declared as an unsafe mechanical system. Use of a mechanical system regulated by this code constituting a hazard to health, safety or welfare by reason of inadequate maintenance, dilapidation, fire hazard, disaster, damage or abandonment is hereby declared an unsafe use. Such unsafe equipment and appliances are hereby declared to be a public nuisance and shall be abated by repair, rehabilitation, demolition or removal.

108.7.1 Authority to condemn mechanical systems. Whenever the code official determines that any mechanical system, or portion thereof, regulated by this code has become hazardous to life, health, property, or has become insanitary, the code official shall order in writing that such system either be removed or restored to a safe condition. A time limit for compliance with such order shall be specified in the written notice. A person shall not use or maintain a defective mechanical system after receiving such notice.

When such mechanical system is to be disconnected, written notice as prescribed in Section 108.2 shall be given. In cases of immediate danger to life or property, such disconnection shall be made immediately without such notice.

108.7.2 Authority to order disconnection of energy sources. The code official shall have the authority to order disconnection of energy sources supplied to a building, structure or mechanical system regulated by this code, when it is determined that the mechanical system or any portion thereof has become hazardous or unsafe. Written notice of such order to disconnect service and the causes therefor shall be given within 24 hours to the owner and occupant of such building, structure or premises, provided, however, that in cases of immediate danger to life or property, such disconnection shall be made immediately without such notice. Where energy sources are provided by a public utility, the code official shall immediately notify the serving utility in writing of the issuance of such order to disconnect.

108.7.3 Connection after order to disconnect. A person shall not make energy source connections to mechanical systems regulated by this code which have been disconnected or ordered to be disconnected by the code official, or the use of which has been ordered to be discontinued by the code official until the code official authorizes the reconnection and use of such mechanical systems.

When a mechanical system is maintained in violation of this code, and in violation of a notice issued pursuant to the provisions of this section, the code official shall institute appropriate action to prevent, restrain, correct or abate the violation.

SECTION 109
MEANS OF APPEAL

109.1 Application for appeal. A person shall have the right to appeal a decision of the code official to the board of appeals. An application for appeal shall be based on a claim that the true intent of this code or the rules legally adopted thereunder have been incorrectly interpreted, the provisions of this code do not fully apply, or an equally good or better form of construction is proposed. The application shall be filed on a form obtained from the code official within 20 days after the notice was served.

109.1.1 Limitation of authority. The board of appeals shall have no authority relative to interpretation of the administration of this code nor shall such board be empowered to waive requirements of this code.

109.2 Membership of board. The board of appeals shall consist of five members appointed by the chief appointing authority as follows: one for five years; one for four years; one for three years; one for two years; and one for one year. Thereafter, each new member shall serve for five years or until a successor has been appointed.

109.2.1 Qualifications. The board of appeals shall consist of five individuals, one from each of the following professions or disciplines.

1. Registered design professional who is a registered architect; or a builder or superintendent of building construction with at least ten years' experience, five of which shall have been in responsible charge of work.

2. Registered design professional with structural engineering or architectural experience.

3. Registered design professional with mechanical and plumbing engineering experience; or a mechanical contractor with at least ten years' experience, five of which shall have been in responsible charge of work.

4. Registered design professional with electrical engineering experience; or an electrical contractor with at least ten years' experience, five of which shall have been in responsible charge of work.

5. Registered design professional with fire protection engineering experience; or a fire protection contractor with at least ten years' experience, five of which shall have been in responsible charge of work.

109.2.2 Alternate members. The chief appointing authority shall appoint two alternate members who shall be called by the board chairman to hear appeals during the absence or disqualification of a member. Alternate members shall possess the qualifications required for board membership and shall be appointed for five years, or until a successor has been appointed.

109.2.3 Chairman. The board shall annually select one of its members to serve as chairman.

109.2.4 Disqualification of member. A member shall not hear an appeal in which that member has a personal, professional or financial interest.

109.2.5 Secretary. The chief administrative officer shall designate a qualified clerk to serve as secretary to the board. The secretary shall file a detailed record of all proceedings in the office of the chief administrative officer.

109.2.6 Compensation of members. Compensation of members shall be determined by law.

109.3 Notice of meeting. The board shall meet upon notice from the chairman, within ten days of the filing of an appeal, or at stated periodic meetings.

109.4 Open hearing. All hearings before the board shall be open to the public. The appellant, the appellant's representative, the code official and any person whose interests are affected shall be given an opportunity to be heard.

109.4.1 Procedure. The board shall adopt and make available to the public through the secretary procedures under which a hearing will be conducted. The procedures shall not require compliance with strict rules of evidence, but shall mandate that only relevant information be received.

109.5 Postponed hearing. When five members are not present to hear an appeal, either the appellant or the appellant's representative shall have the right to request a postponement of the hearing.

109.6 Board decision. The board shall modify or reverse the decision of the code official by a concurring vote of three members.

109.6.1 Resolution. The decision of the board shall be by resolution. Certified copies shall be furnished to the appellant and to the code official.

109.6.2 Administration. The code official shall take immediate action in accordance with the decision of the board.

109.7 Court review. Any person, whether or not a previous party of the appeal, shall have the right to apply to the appropriate court for a writ of certiorari to correct errors of law. Application for review shall be made in the manner and time required by law following the filing of the decision in the office of the chief administrative officer.

CHAPTER 2
DEFINITIONS

201.1 Scope. Unless otherwise expressly stated, the following words and terms shall, for the purposes of this code, have the meanings indicated in this chapter.

201.2 Interchangeability. Words used in the present tense include the future; words in the masculine gender include the feminine and neuter; the singular number includes the plural and the plural, the singular.

201.3 Terms defined in other codes. Where terms are not defined in this code and are defined in the *International Building Code*, ICC *Electrical Code, International Fire Code, International Fuel Gas Code* or the *International Plumbing Code*, such terms shall have meanings ascribed to them as in those codes.

201.4 Terms not defined. Where terms are not defined through the methods authorized by this section, such terms shall have ordinarily accepted meanings such as the context implies.

SECTION 202
GENERAL DEFINITIONS

ABRASIVE MATERIALS. Moderately abrasive particulate in high concentrations, and highly abrasive particulate in moderate and high concentrations, such as alumina, bauxite, iron silicate, sand and slag.

ABSORPTION SYSTEM. A refrigerating system in which refrigerant is pressurized by pumping a chemical solution of refrigerant in absorbent, and then separated by the addition of heat in a generator, condensed (to reject heat), expanded, evaporated (to provide refrigeration), and reabsorbed in an absorber to repeat the cycle; the system may be single or multiple effect, the latter using multiple stages or internally cascaded use of heat to improve efficiency.

ACCESS (TO). That which enables a device, appliance or equipment to be reached by ready access or by a means that first requires the removal or movement of a panel, door or similar obstruction [see also "Ready access (to)"].

AIR. All air supplied to mechanical equipment and appliances for combustion, ventilation, cooling, etc. Standard air is air at standard temperature and pressure, namely, 70°F (21°C) and 29.92 inches of mercury (101.3 kPa).

AIR CONDITIONING . The treatment of air so as to control simultaneously the temperature, humidity, cleanness and distribution of the air to meet the requirements of a conditioned space.

AIR-CONDITIONING SYSTEM. A system that consists of heat exchangers, blowers, filters, supply, exhaust and return ducts, and shall include any apparatus installed in connection therewith.

AIR DISTRIBUTION SYSTEM. Any system of ducts, plenums and air-handling equipment that circulates air within a space or spaces and includes systems made up of one or more air-handling units.

AIR, EXHAUST. Air being removed from any space, appliance or piece of equipment and conveyed directly to the atmosphere by means of openings or ducts.

AIR-HANDLING UNIT. A blower or fan used for the purpose of distributing supply air to a room, space or area.

AIR, MAKEUP. Air that is provided to replace air being exhausted.

ALTERATION. A change in a mechanical system that involves an extension, addition or change to the arrangement, type or purpose of the original installation.

APPLIANCE. A device or apparatus that is manufactured and designed to utilize energy and for which this code provides specific requirements.

APPLIANCE, EXISTING . Any appliance regulated by this code which was legally installed prior to the effective date of this code, or for which a permit to install has been issued.

APPLIANCE TYPE.

> **High-heat appliance.** Any appliance in which the products of combustion at the point of entrance to the flue under normal operating conditions have a temperature greater than 2,000°F (1093°C).

> **Low-heat appliance (residential appliance).** Any appliance in which the products of combustion at the point of entrance to the flue under normal operating conditions have a temperature of 1,000°F (538°C) or less.

> **Medium-heat appliance.** Any appliance in which the products of combustion at the point of entrance to the flue under normal operating conditions have a temperature of more than 1,000°F (538°C), but not greater than 2,000°F (1093°C).

APPLIANCE, VENTED. An appliance designed and installed in such a manner that all of the products of combustion are conveyed directly from the appliance to the outside atmosphere through an approved chimney or vent system.

APPROVED. Approved by the code official or other authority having jurisdiction.

APPROVED AGENCY. An established and recognized agency that is approved by the code official and regularly engaged in conducting tests or furnishing inspection services.

AUTOMATIC BOILER. Any class of boiler that is equipped with the controls and limit devices specified in Chapter 10.

BATHROOM. A room containing a bathtub, shower, spa or similar bathing fixture.

BOILER. A closed heating appliance intended to supply hot water or steam for space heating, processing or power purposes. Low-pressure boilers operate at pressures less than or equal to 15 pounds per square inch (psi) (103 kPa) for steam and 160 psi (1103 kPa) for water. High-pressure boilers operate at pressures exceeding those pressures.

BOILER ROOM. A room primarily utilized for the installation of a boiler.

BRAZED JOINT. A gas-tight joint obtained by the joining of metal parts with metallic mixtures or alloys which melt at a temperature above 1,000°F (538°C), but lower than the melting temperature of the parts to be joined.

BRAZING. A metal joining process wherein coalescence is produced by the use of a nonferrous filler metal having a melting point above 1,000°F (538°C), but lower than that of the base metal being joined. The filler material is distributed between the closely fitted surfaces of the joint by capillary attraction.

Btu. Abbreviation for British thermal unit, which is the quantity of heat required to raise the temperature of 1 pound (454 g) of water 1°F (0.56°C) (1 Btu = 1055 J).

BUILDING. Any structure occupied or intended for supporting or sheltering any occupancy.

CHIMNEY. A primarily vertical structure containing one or more flues, for the purpose of carrying gaseous products of combustion and air from a fuel-burning appliance to the outside atmosphere.

Factory-built chimney. A listed and labeled chimney composed of factory-made components, assembled in the field in accordance with manufacturer's instructions and the conditions of the listing.

Masonry chimney. A field-constructed chimney composed of solid masonry units, bricks, stones or concrete.

Metal chimney. A field-constructed chimney of metal.

CHIMNEY CONNECTOR. A pipe that connects a fuel-burning appliance to a chimney.

CLEARANCE. The minimum distance through air measured between the heat-producing surface of the mechanical appliance, device or equipment and the surface of the combustible material or assembly.

CLOSED COMBUSTION SOLID-FUEL-BURNING APPLIANCE. A heat-producing appliance that employs a combustion chamber that has no openings other than the flue collar, fuel charging door and adjustable openings provided to control the amount of combustion air that enters the combustion chamber.

CLOTHES DRYER. An appliance used to dry wet laundry by means of heat. Dryer classifications are as follows:

Type 1. Factory-built package, multiple production. Primarily used in family living environment. Usually the smallest unit physically and in function output.

Type 2. Factory-built package, multiple production. Used in business with direct intercourse of the function with the public. Not designed for use in individual family living environment.

CODE. These regulations, subsequent amendments thereto, or any emergency rule or regulation that the administrative authority having jurisdiction has lawfully adopted.

CODE OFFICIAL. The officer or other designated authority charged with the administration and enforcement of this code, or a duly authorized representative.

COMBUSTIBLE ASSEMBLY. Wall, floor, ceiling or other assembly constructed of one or more component materials that are not defined as noncombustible.

COMBUSTIBLE LIQUIDS. Any liquids having a flash point at or above 100°F (38°C), and that are divided into the following classifications:

Class II. Liquids having flash points at or above 100°F (38°C) and below 140°F (60°C).

Class IIIA. Liquids having flash points at or above 140°F (60°C) and below 200°F (93°C).

Class IIIB. Liquids having flash points at or above 200°F (93°C).

COMBUSTIBLE MATERIAL. Any material not defined as noncombustible.

COMBUSTION. In the context of this code, refers to the rapid oxidation of fuel accompanied by the production of heat or heat and light.

COMBUSTION AIR. Air necessary for complete combustion of a fuel, including theoretical air and excess air.

COMBUSTION CHAMBER. The portion of an appliance within which combustion occurs.

COMBUSTION PRODUCTS. Constituents resulting from the combustion of a fuel with the oxygen of the air, including the inert gases, but excluding excess air.

COMMERCIAL COOKING RECIRCULATING SYSTEM. Self-contained system consisting of the exhaust hood, the cooking equipment, the filters, and the fire suppression system. The system is designed to capture cooking vapors and residues generated from commercial cooking equipment. The system removes contaminants from the exhaust air and recirculates the air to the space from which it was withdrawn.

COMMERCIAL COOKING APPLIANCES. Appliances used in a commercial food service establishment for heating or cooking food and which produce grease vapors, steam, fumes, smoke or odors that are required to be removed through a local exhaust ventilation system. Such appliances include deep fat fryers; upright broilers; griddles; broilers; steam-jacketed kettles; hot-top ranges; under-fired broilers (charbroilers); ovens; barbecues; rotisseries; and similar appliances. For the purpose of this definition, a food service establishment shall include any building or a portion thereof used for the preparation and serving of food.

COMMERCIAL KITCHEN HOODS.

Backshelf Hood. A backshelf hood is also referred to as a low-proximity hood, or as a sidewall hood where wall mounted. It's front lower lip is low over the appliance(s) and is "set back" from the front of the appliance(s). It is always closed to the rear of the appliances by a panel where free-standing, or by a panel or wall where wall mounted, and its height above the cooking surface varies. (This style of hood can be constructed with partial end panels to increase its effectiveness in capturing the effluent generated by the cooking operation).

Double Island Canopy Hood. A double island canopy hood is placed over back to back appliances or appliance lines. It is open on all sides and overhangs both fronts and the sides of the appliance(s). It could have a wall panel between the backs of the appliances. (The fact that exhaust air is drawn from both sides of the double canopy to meet in the center causes each side of this hood to emulate a wall canopy hood, and thus it functions much the same with or without an actual wall panel between the backs of the appliances).

Eyebrow Hood. An eyebrow hood is mounted directly to the face of an appliance, such as an oven and dishwasher, above the opening(s) or door(s) from which effluent is emitted, extending past the sides and overhanging the front of the opening to capture the effluent.

Pass-over Hood. A pass-over hood is a free-standing form of a backshelf hood constructed low enough to pass food over the top.

Single Island Canopy Hood. A single island canopy hood is placed over a single appliance or appliance line. It is open on all sides and overhangs the front, rear, and sides of the appliance(s). A single island canopy is more susceptible to cross drafts and requires a greater exhaust air flow than an equivalent sized wall-mounted canopy to capture and contain effluent generated by the cooking operation(s).

Wall Canopy Hood. A wall canopy exhaust hood is mounted against a wall above a single appliance or line of appliance(s), or it could be free-standing with a back panel from the rear of the appliances to the hood. It overhangs the front and sides of the appliance(s) on all open sides.

(The wall acts as a back panel, forcing the makeup air to be drawn across the front of the cooking equipment, thus increasing the effectiveness of the hood to capture and contain effluent generated by the cooking operation(s).

COMPENSATING HOODS. Compensating hoods are those having integral (built-in) makeup air supply. The makeup air supply for such hoods is generally supplied from: short-circuit flow from inside the hood, air curtain flow from the bottom of the front face, and front face discharge from the outside front wall of the hood. The compensating makeup airflow can also be supplied from the rear or side of the hood, or the rear, front, or sides of the cooking equipment. The makeup airflow can be one or a combination of methods.

COMPRESSOR. A specific machine, with or without accessories, for compressing a gas.

COMPRESSOR, POSITIVE DISPLACEMENT. A compressor in which increase in pressure is attained by changing the internal volume of the compression chamber.

COMPRESSOR UNIT. A compressor with its prime mover and accessories.

CONCEALED LOCATION. A location that cannot be accessed without damaging permanent parts of the building structure or finish surface. Spaces above, below or behind readily removable panels or doors shall not be considered as concealed.

CONDENSATE. The liquid that condenses from a gas (including flue gas) caused by a reduction in temperature.

CONDENSER. A heat exchanger designed to liquefy refrigerant vapor by removal of heat.

CONDENSING UNIT. A specific refrigerating machine combination for a given refrigerant, consisting of one or more power-driven compressors, condensers, liquid receivers (when required), and the regularly furnished accessories.

CONDITIONED SPACE. An area, room or space being heated or cooled by any equipment or appliance.

CONFINED SPACES. A space having a volume less than 50 cubic feet per 1,000 British thermal units per hour (Btu/h) (4.8 m³/kW) of the aggregate input rating of all appliances installed in that space.

CONSTRUCTION DOCUMENTS. All of the written, graphic and pictorial documents prepared or assembled for describing the design, location and physical characteristics of the elements of the project necessary for obtaining a building permit. The construction drawings shall be drawn to an appropriate scale.

CONTROL. A manual or automatic device designed to regulate the gas, air, water or electrical supply to, or operation of, a mechanical system.

CONVERSION BURNER. A burner designed to supply gaseous fuel to an appliance originally designed to utilize another fuel.

COOKING APPLIANCE. See "Commercial cooking appliances."

DAMPER. A manually or automatically controlled device to regulate draft or the rate of flow of air or combustion gases.

> **Volume damper.** A device that, when installed, will restrict, retard or direct the flow of air in a duct, or the products of combustion in a heat-producing appliance, its vent connector, vent or chimney therefrom.

DESIGN FLOOD ELEVATION. The elevation of the "design flood," including wave height, relative to the datum specified on the community's legally designated flood hazard area map.

DESIGN WORKING PRESSURE. The maximum allowable working pressure for which a specific part of a system is designed.

DIRECT REFRIGERATION SYSTEM. A system in which the evaporator or condenser of the refrigerating system is in direct contact with the air or other substances to be cooled or heated.

DIRECT-VENT APPLIANCES. Appliances that are constructed and installed so that all air for combustion is derived from the outside atmosphere and all flue gases are discharged to the outside atmosphere.

DRAFT. The pressure difference existing between the appliance or any component part and the atmosphere, that causes a continuous flow of air and products of combustion through the gas passages of the appliance to the atmosphere.

Induced draft. The pressure difference created by the action of a fan, blower or ejector, that is located between the appliance and the chimney or vent termination.

Natural draft. The pressure difference created by a vent or chimney because of its height, and the temperature difference between the flue gases and the atmosphere.

DRIP. The container placed at a low point in a system of piping to collect condensate and from which the condensate is removable.

DRY CLEANING SYSTEMS. Dry cleaning plants or systems are classified as follows:

Type I. Those systems using Class I flammable liquid solvents having a flash point below 100°F (38°C).

Type II. Those systems using Class II combustible liquid solvents having a flash point at or above 100°F (38°C) and below 140°F (60°C).

Type III. Those systems using Class III combustible liquid solvents having a flash point at or above 140°F (60°C).

Types IV and V. Those systems using Class IV nonflammable liquid solvents.

DUCT. A tube or conduit utilized for conveying air. The air passages of self-contained systems are not to be construed as air ducts.

DUCT FURNACE. A warm-air furnace normally installed in an air distribution duct to supply warm air for heating. This definition shall apply only to a warm-air heating appliance that, for air circulation, depends on a blower not furnished as part of the furnace.

DUCT SYSTEM. A continuous passageway for the transmission of air that, in addition to ducts, includes duct fittings, dampers, plenums, fans and accessory air-handling equipment and appliances.

DWELLING. A building or portion thereof that contains not more than two dwelling units.

DWELLING UNIT. A single unit providing complete, independent living facilities for one or more persons, including permanent provisions for living, sleeping, eating, cooking and sanitation.

ELECTRIC HEATING APPLIANCE. An appliance that produces heat energy to create a warm environment by the application of electric power to resistance elements, refrigerant compressors or dissimilar material junctions.

ENERGY RECOVERY VENTILATION SYSTEM. Systems that employ air-to-air heat exchangers to recover energy from or reject energy to exhaust air for the purpose of pre-heating, pre-cooling, humidifying or dehumidifying outdoor ventilation air prior to supplying such air to a space, either directly or as part of an HVAC system.

EQUIPMENT. All piping, ducts, vents, control devices and other components of systems other than appliances which are permanently installed and integrated to provide control of environmental conditions for buildings. This definition shall also include other systems specifically regulated in this code.

EQUIPMENT, EXISTING . Any equipment regulated by this code which was legally installed prior to the effective date of this code, or for which a permit to install has been issued.

EVAPORATIVE COOLER. A device used for reducing the sensible heat of air for cooling by the process of evaporation of water into an airstream.

EVAPORATIVE COOLING SYSTEM. The equipment and appliances intended or installed for the purpose of environmental cooling by an evaporative cooler from which the conditioned air is distributed through ducts or plenums to the conditioned area.

EVAPORATOR. That part of the system in which liquid refrigerant is vaporized to produce refrigeration.

EXCESS AIR. The amount of air provided in addition to theoretical air to achieve complete combustion of a fuel, thereby preventing the formation of dangerous products of combustion.

EXHAUST SYSTEM. An assembly of connected ducts, plenums, fittings, registers, grilles and hoods through which air is conducted from the space or spaces and exhausted to the outside atmosphere.

EXTRA-HEAVY DUTY COOKING APPLIANCE. Extra-heavy duty cooking appliances include appliances utilizing solid fuel such as wood, charcoal, briquettes, and mesquite as the primary source of heat for cooking.

FIREPLACE. An assembly consisting of a hearth and fire chamber of noncombustible material and provided with a chimney, for use with solid fuels.

Factory-built fireplace. A listed and labeled fireplace and chimney system composed of factory-made components, and assembled in the field in accordance with manufacturer's instructions and the conditions of the listing.

Masonry fireplace. A field-constructed fireplace composed of solid masonry units, bricks, stones or concrete.

FIREPLACE STOVE. A free-standing chimney-connected solid-fuel-burning heater, designed to be operated with the fire chamber doors in either the open or closed position.

FLAME SAFEGUARD. A device that will automatically shut off the fuel supply to a main burner or group of burners when the means of ignition of such burners becomes inoperative, and when flame failure occurs on the burner or group of burners.

FLAME SPREAD INDEX. The numerical value assigned to a material tested in accordance with ASTM E 84.

FLAMMABILITY CLASSIFICATION. Refrigerants shall be assigned to one of the three classes—1, 2 or 3—in accordance with ASHRAE 34. For Classes 2 and 3, the heat of combustion shall be calculated assuming that combustion products are in the gas phase and in their most stable state.

Class 1. Refrigerants that do not show flame propagation when tested in air at 14.7 psia (101 kPa) and 70°F (21°C).

Class 2. Refrigerants having a lower flammability limit (LFL) of more than 0.00625 pound per cubic foot (0.10

kg/m³) at 70°F (21°C) and 14.7 psia (101 kPa) and a heat of combustion of less than 8,174 Btu/lb (19 000 kJ/kg).

Class 3. Refrigerants that are highly flammable, having a LFL of less than or equal to 0.00625 pound per cubic foot (0.10 kg/m³) at 70°F (21°C) and 14.7 psia (101 kPa) or a heat of combustion greater than or equal to 8,174 Btu/lb (19 000 kJ/kg).

FLAMMABLE LIQUIDS. Any liquid that has a flash point below 100°F (38°C), and has a vapor pressure not exceeding 40 psia (276 kPa) at 100°F (38°C). Flammable liquids shall be known as Class I liquids and shall be divided into the following classifications:

Class IA. Liquids having a flash point below 73°F (23°C) and a boiling point below 100°F (38°C).

Class IB. Liquids having a flash point below 73°F (23°C) and a boiling point at or above 100°F (38°C).

Class IC. Liquids having a flash point at or above 73°F (23°C) and below 100°F (38°C).

FLAMMABLE VAPOR OR FUMES. Mixtures of gases in air at concentrations equal to or greater than the LFL and less than or equal to the upper flammability limit (UFL).

FLASH POINT. The minimum temperature corrected to a pressure of 14.7 psia (101 kPa) at which the application of a test flame causes the vapors of a portion of the sample to ignite under the conditions specified by the test procedures and apparatus. The flash point of a liquid shall be determined in accordance with ASTM D 56, ASTM D 93 or ASTM D 3278.

FLOOR AREA, NET. The actual occupied area, not including unoccupied accessory areas or thicknesses of walls.

FLOOR FURNACE. A completely self-contained furnace suspended from the floor of the space being heated, taking air for combustion from outside such space and with means for observing flames and lighting the appliance from such space.

FLUE. A passageway within a chimney or vent through which gaseous combustion products pass.

FLUE CONNECTION (BREECHING). A passage for conducting the products of combustion from a fuel-fired appliance to the vent or chimney (see also "Chimney connector" and "Vent connector").

FLUE GASES. Products of combustion and excess air.

FLUE LINER (LINING). A system or material used to form the inside surface of a flue in a chimney or vent, for the purpose of protecting the surrounding structure from the effects of combustion products and conveying combustion products without leakage to the atmosphere.

FUEL GAS. A natural gas, manufactured gas, liquefied petroleum gas or a mixture of these.

FUEL OIL. Kerosene or any hydrocarbon oil having a flash point not less than 100°F (38°C).

FUEL-OIL PIPING SYSTEM. A closed piping system that connects a combustible liquid from a source of supply to a fuel-oil-burning appliance.

FURNACE. A completely self-contained heating unit that is designed to supply heated air to spaces remote from or adjacent to the appliance location.

FURNACE ROOM. A room primarily utilized for the installation of fuel-burning space-heating and water-heating appliances other than boilers (see also "Boiler room").

FUSIBLE PLUG. A device arranged to relieve pressure by operation of a fusible member at a predetermined temperature.

GROUND SOURCE HEAT PUMP LOOP SYSTEM. Piping buried in horizontal or vertical excavations or placed in a body of water for the purpose of transporting heat transfer liquid to and from a heat pump. Included in this definition are closed loop systems in which the liquid is recirculated and open loop systems in which the liquid is drawn from a well or other source.

HAZARDOUS LOCATION. Any location considered to be a fire hazard for flammable vapors, dust, combustible fibers or other highly combustible substances. The location is not necessarily categorized in the *International Building Code* as a high-hazard use group classification.

HEAT EXCHANGER. A device that transfers heat from one medium to another.

HEAT PUMP. A refrigeration system that extracts heat from one substance and transfers it to another portion of the same substance or to a second substance at a higher temperature for a beneficial purpose.

HEAT TRANSFER LIQUID. The operating or thermal storage liquid in a mechanical system, including water or other liquid base, and additives at the concentration present under operating conditions used to move heat from one location to another. Refrigerants are not included as heat transfer liquids.

HEAVY-DUTY COOKING APPLIANCE. Heavy-duty cooking appliances include electric under-fired broilers, electric chain (conveyor) broilers, gas under-fired broilers, gas chain (conveyor) broilers, gas open-burner ranges (with or without oven), electric and gas wok ranges, and electric and gas over-fired (upright) broilers and salamanders.

HIGH-PROBABILITY SYSTEMS. A refrigeration system in which the basic design or the location of components is such that a leakage of refrigerant from a failed connection, seal or component will enter an occupancy classified area, other than the machinery room.

HIGH-SIDE PRESSURE. The parts of a refrigerating system subject to condenser pressure.

HOOD. An air-intake device used to capture by entrapment, impingement, adhesion or similar means, grease and similar contaminants before they enter a duct system.

Type I. A kitchen hood for collecting and removing grease vapors and smoke.

Type II. A general kitchen hood for collecting and removing steam, vapor, heat and odors.

HYDROGEN GENERATING APPLIANCE. A self-contained package or factory-matched packages of integrated systems for generating gaseous hydrogen. Hydrogen generating appliances utilize electrolysis, reformation, chemical, or other processes to generate hydrogen.

IGNITION SOURCE. A flame, spark or hot surface capable of igniting flammable vapors or fumes. Such sources include appliance burners, burner ignitors and electrical switching devices.

IMMEDIATELY DANGEROUS TO LIFE OR HEALTH (IDLH). The concentration of airborne contaminants that poses a threat of death, immediate or delayed permanent adverse health effects, or effects that could prevent escape from such an environment. This contaminant concentration level is established by the National Institute of Occupational Safety and Health (NIOSH) based on both toxicity and flammability. It is generally expressed in parts per million by volume (ppm v/v) or milligrams per cubic meter (mg/m^3).

INDIRECT REFRIGERATION SYSTEM. A system in which a secondary coolant cooled or heated by the refrigerating system is circulated to the air or other substance to be cooled or heated. Indirect systems are distinguished by the method of application shown below:

    Closed system. A system in which a secondary fluid is either cooled or heated by the refrigerating system and then circulated within a closed circuit in indirect contact with the air or other substance to be cooled or heated.

    Double-indirect open-spray system. A system in which the secondary substance for an indirect open-spray system is heated or cooled by an intermediate coolant circulated from a second enclosure.

    Open-spray system. A system in which a secondary coolant is cooled or heated by the refrigerating system and then circulated in direct contact with the air or other substance to be cooled or heated.

    Vented closed system. A system in which a secondary coolant is cooled or heated by the refrigerating system and then passed through a closed circuit in the air or other substance to be cooled or heated, except that the evaporator or condenser is placed in an open or appropriately vented tank.

JOINT, FLANGED. A joint made by bolting together a pair of flanged ends.

JOINT, FLARED. A metal-to-metal compression joint in which a conical spread is made on the end of a tube that is compressed by a flare nut against a mating flare.

JOINT, MECHANICAL. A general form of gas-tight joints obtained by the joining of metal parts through a positive-holding mechanical construction, such as flanged joint, screwed joint or flared joint.

JOINT, PLASTIC ADHESIVE. A joint made in thermoset plastic piping by the use of an adhesive substance which forms a continuous bond between the mating surfaces without dissolving either one of them.

JOINT, PLASTIC HEAT FUSION. A joint made in thermoplastic piping by heating the parts sufficiently to permit fusion of the materials when the parts are pressed together.

JOINT, PLASTIC SOLVENT CEMENT. A joint made in thermoplastic piping by the use of a solvent or solvent cement which forms a continuous bond between the mating surfaces.

JOINT, SOLDERED. A gas-tight joint obtained by the joining of metal parts with metallic mixtures of alloys which melt at temperatures between 400°F (204°C) and 1,000°F (538°C).

JOINT, WELDED. A gas-tight joint obtained by the joining of metal parts in molten state.

LABELED. Devices, equipment, appliances or materials to which have been affixed a label, seal, symbol or other identifying mark of a nationally recognized testing laboratory, inspection agency or other organization concerned with product evaluation that maintains periodic inspection of the production of the above-labeled items and by whose label the manufacturer attests to compliance with applicable nationally recognized standards.

LIGHT-DUTY COOKING APPLIANCE. Light-duty cooking appliances include gas and electric ovens (including standard, bake, roasting, revolving, retherm, convection, combination convection/steamer, conveyor, deck or deck-style pizza, and pastry), electric and gas steam-jacketed kettles, electric and gas compartment steamers (both pressure and atmospheric) and electric and gas cheesemelters.

LIMIT CONTROL. A device responsive to changes in pressure, temperature or level for turning on, shutting off or throttling the gas supply to an appliance.

LIMITED CHARGE SYSTEM. A system in which, with the compressor idle, the design pressure will not be exceeded when the refrigerant charge has completely evaporated.

LISTED. Equipment, appliances or materials included in a list published by a nationally recognized testing laboratory, inspection agency or other organization concerned with product evaluation that maintains periodic inspection of production of listed equipment, appliances or materials, and whose listing states either that the equipment, appliances or material meets nationally recognized standards or has been tested and found suitable for use in a specified manner. Not all testing laboratories, inspection agencies and other organizations concerned with product evaluation use the same means for identifying listed equipment, appliances or materials. Some do not recognize equipment, appliances or materials as listed unless they are also labeled. The authority having jurisdiction shall utilize the system employed by the listing organization to identify a listed product.

LIVING SPACE. Space within a dwelling unit utilized for living, sleeping, eating, cooking, bathing, washing and sanitation purposes.

LOWER EXPLOSIVE LIMIT (LEL). See "LFL."

LOWER FLAMMABILITY LIMIT (LFL). The minimum concentration of refrigerant that is capable of propagating a flame through a homogeneous mixture of refrigerant and air.

LOW-PRESSURE HOT-WATER-HEATING BOILER. A boiler furnishing hot water at pressures not exceeding 160 psi (1103 kPa) and at temperatures not exceeding 250°F (121°C).

LOW-PRESSURE STEAM-HEATING BOILER. A boiler furnishing steam at pressures not exceeding 15 psi (103 kPa).

LOW-PROBABILITY SYSTEMS. A refrigeration system in which the basic design or the location of components is such that a leakage of refrigerant from a failed connection, seal or component will not enter an occupancy-classified area, other than the machinery room.

LOW-SIDE PRESSURE. The parts of a refrigerating system subject to evaporator pressure.

MACHINERY ROOM. A room meeting prescribed safety requirements and in which refrigeration systems or components thereof are located (see Sections 1105 and 1106).

MECHANICAL DRAFT SYSTEM. A venting system designed to remove flue or vent gases by mechanical means, that consists of an induced-draft portion under nonpositive static pressure or a forced-draft portion under positive static pressure.

> **Forced-draft venting system.** A portion of a venting system using a fan or other mechanical means to cause the removal of flue or vent gases under positive static pressure.
>
> **Induced-draft venting system.** A portion of a venting system using a fan or other mechanical means to cause the removal of flue or vent gases under nonpositive static vent pressure.
>
> **Power venting system.** A portion of a venting system using a fan or other mechanical means to cause the removal of flue or vent gases under positive static vent pressure.

MECHANICAL EQUIPMENT/APPLIANCE ROOM. A room or space in which nonfuel-fired mechanical equipment and appliances are located.

MECHANICAL EXHAUST SYSTEM. A system for removing air from a room or space by mechanical means.

MECHANICAL JOINT. A connection between pipes, fittings, or pipes and fittings, which is neither screwed, caulked, threaded, soldered, solvent cemented, brazed nor welded. Also, a joint in which compression is applied along the centerline of the pieces being joined. Some joints are part of a coupling, fitting or adapter.

MECHANICAL SYSTEM. A system specifically addressed and regulated in this code and composed of components, devices, appliances and equipment.

MEDIUM-DUTY COOKING APPLIANCE. Medium-duty cooking appliances include electric discrete element ranges (with or without oven), electric and gas hot-top ranges, electric and gas griddles, electric and gas double-sided griddles, electric and gas fryers (including open deep fat fryers, donut fryers, kettle fryers, and pressure fryers), electric and gas pasta cookers, electric and gas conveyor pizza ovens, electric and gas tilting skillets (braising pans) and electric and gas rotisseries.

MODULAR BOILER. A steam or hot-water-heating assembly consisting of a group of individual boilers called modules intended to be installed as a unit with no intervening stop valves. Modules are under one jacket or are individually jacketed. The individual modules shall be limited to a maximum input rating of 400,000 Btu/h (117 228 W) gas, 3 gallons per hour (gph) (11.4 L/h) oil, or 115 kW (electric).

NATURAL DRAFT SYSTEM. A venting system designed to remove flue or vent gases under nonpositive static vent pressure entirely by natural draft.

NATURAL VENTILATION. The movement of air into and out of a space through intentionally provided openings, such as windows and doors, or through nonpowered ventilators.

NONABRASIVE/ABRASIVE MATERIALS. Nonabrasive particulate in high concentrations, moderately abrasive particulate in low and moderate concentrations, and highly abrasive particulate in low concentrations, such as alfalfa, asphalt, plaster, gypsum and salt.

NONCOMBUSTIBLE MATERIALS. Materials that, when tested in accordance with ASTM E 136, have at least three of four specimens tested meeting all of the following criteria:

1. The recorded temperature of the surface and interior thermocouples shall not at any time during the test rise more than 54°F (30°C) above the furnace temperature at the beginning of the test.

2. There shall not be flaming from the specimen after the first 30 seconds.

3. If the weight loss of the specimen during testing exceeds 50 percent, the recorded temperature of the surface and interior thermocouples shall not at any time during the test rise above the furnace air temperature at the beginning of the test, and there shall not be flaming of the specimen.

OCCUPANCY. The purpose for which a building, or portion thereof, is utilized or occupied.

OFFSET (VENT). A combination of approved bends that make two changes in direction bringing one section of the vent out of line but into a line parallel with the other section.

OUTDOOR AIR. Air taken from the outdoors, and therefore not previously circulated through the system.

OUTDOOR OPENING. A door, window, louver or skylight openable to the outside atmosphere.

OUTLET. A threaded connection or bolted flange in a piping system to which a gas-burning appliance is attached.

PANEL HEATING. A method of radiant space heating in which heat is supplied by large heated areas of room surfaces. The heating element usually consists of warm water piping, warm air ducts, or electrical resistance elements embedded in or located behind ceiling, wall or floor surfaces.

PELLET FUEL-BURNING APPLIANCE. A closed-combustion, vented appliance equipped with a fuel-feed mechanism for burning processed pellets of solid fuel of a specified size and composition.

PIPING. Where used in this code, "piping" refers to either pipe or tubing, or both.

> **Pipe.** A rigid conduit of iron, steel, copper, brass or plastic.
>
> **Tubing.** Semirigid conduit of copper, aluminum, plastic or steel.

PLASTIC, THERMOPLASTIC. A plastic that is capable of being repeatedly softened by increase of temperature and hardened by decrease of temperature.

PLASTIC, THERMOSETTING. A plastic that is capable of being changed into a substantially infusible or insoluble product when cured under application of heat or chemical means.

PLENUM. An enclosed portion of the building structure, other than an occupiable space being conditioned, that is designed to allow air movement, and thereby serve as part of an air distribution system.

PORTABLE FUEL CELL APPLIANCE. A fuel cell generator of electricity, which is not fixed in place. A portable fuel cell appliance utilizes a cord and plug connection to a grid-isolated load and has an integral fuel supply.

POWER BOILER. See "Boiler."

PREMISES. A lot, plot or parcel of land, including any structure thereon.

PRESSURE, FIELD TEST. A test performed in the field to prove system tightness.

PRESSURE-LIMITING DEVICE. A pressure-responsive mechanism designed to stop automatically the operation of the pressure-imposing element at a predetermined pressure.

PRESSURE RELIEF DEVICE. A pressure-actuated valve or rupture member designed to relieve excessive pressure automatically.

PRESSURE RELIEF VALVE. A pressure-actuated valve held closed by a spring or other means and designed to relieve pressure automatically in excess of the device's setting.

PRESSURE VESSELS. Closed containers, tanks or vessels that are designed to contain liquids or gases, or both, under pressure.

PRESSURE VESSELS—REFRIGERANT. Any refrigerant-containing receptacle in a refrigerating system. This does not include evaporators where each separate section does not exceed 0.5 cubic foot (0.014 m³) of refrigerant-containing volume, regardless of the maximum inside dimensions, evaporator coils, controls, headers, pumps and piping.

PROTECTIVE ASSEMBLY (REDUCED CLEARANCE). Any noncombustible assembly that is labeled or constructed in accordance with Table 308.6 and is placed between combustible materials or assemblies and mechanical appliances, devices or equipment, for the purpose of reducing required airspace clearances. Protective assemblies attached directly to a combustible assembly shall not be considered as part of that combustible assembly.

PURGE. To clear of air, water or other foreign substances.

QUICK-OPENING VALVE. A valve that opens completely by fast action, either manually or automatically controlled. A valve requiring one-quarter round turn or less is considered to be quick opening.

RADIANT HEATER. A heater designed to transfer heat primarily by direct radiation.

READY ACCESS (TO). That which enables a device, appliance or equipment to be directly reached, without requiring the removal or movement of any panel, door or similar obstruction [see "Access (to)"].

RECEIVER, LIQUID. A vessel permanently connected to a refrigeration system by inlet and outlet pipes for storage of liquid refrigerant.

RECIRCULATED AIR. Air removed from a conditioned space and intended for reuse as supply air.

RECLAIMED REFRIGERANTS. Refrigerants reprocessed to the same specifications as for new refrigerants by means including distillation. Such refrigerants have been chemically analyzed to verify that the specifications have been met. Reclaiming usually implies the use of processes or procedures that are available only at a reprocessing or manufacturing facility.

RECOVERED REFRIGERANTS. Refrigerants removed from a system in any condition without necessarily testing or processing them.

RECYCLED REFRIGERANTS. Refrigerants from which contaminants have been reduced by oil separation, removal of noncondensable gases, and single or multiple passes through devices that reduce moisture, acidity and particulate matter, such as replaceable core filter driers. These procedures usually are performed at the field job site or in a local service shop.

REFRIGERANT. A substance utilized to produce refrigeration by its expansion or vaporization.

REFRIGERANT SAFETY CLASSIFICATIONS. Groupings that indicate the toxicity and flammability classes in accordance with Section 1103.1. The classification group is made up of a letter (A or B) that indicates the toxicity class, followed by a number (1, 2 or 3) that indicates the flammability class. Refrigerant blends are similarly classified, based on the compositions at their worst cases of fractionation, as separately determined for toxicity and flammability. In some cases, the worst case of fractionation is the original formulation.

Flammability. Class 1 indicates refrigerants that do not show flame propagation in air when tested by prescribed methods at specified conditions. Classes 2 and 3 signify refrigerants with "lower flammability" and "higher flammability," respectively; the distinction depends on both the LFL and heat of combustion.

Toxicity. Classes A and B signify refrigerants with "lower toxicity" and "higher toxicity," respectively, based on prescribed measures of chronic (long-term, repeated exposures) toxicity.

REFRIGERATED ROOM OR SPACE. A room or space in which an evaporator or brine coil is located for the purpose of reducing or controlling the temperature within the room or space to below 68°F (20°C).

REFRIGERATING SYSTEM. A combination of interconnected refrigerant-containing parts constituting one closed refrigerant circuit in which a refrigerant is circulated for the purpose of extracting heat.

REFRIGERATION CAPACITY RATING. Expressed as 1 horsepower (0.75 kW), 1 ton or 12,000 Btu/h (3.5 kW), shall all mean the same quantity.

REFRIGERATION MACHINERY ROOM. See "Machinery room."

REFRIGERATION SYSTEM, ABSORPTION. A heat-operated, closed-refrigeration cycle in which a secondary fluid (the absorbent) absorbs a primary fluid (the refrigerant) that has been vaporized in the evaporator.

Direct system. A system in which the evaporator is in direct contact with the material or space refrigerated, or is located in air-circulating passages communicating with such spaces.

Indirect system. A system in which a brine coil cooled by the refrigerant is circulated to the material or space refrigerated, or is utilized to cool the air so circulated. Indirect systems are distinguished by the type or method of application.

REFRIGERATION SYSTEM CLASSIFICATION. Refrigeration systems are classified according to the degree of probability that leaked refrigerant from a failed connection, seal or component will enter an occupied area. The distinction is based on the basic design or location of the components.

REFRIGERATION SYSTEM, MECHANICAL. A combination of interconnected refrigeration-containing parts constituting one closed refrigerant circuit in which a refrigerant is circulated for the purpose of extracting heat and in which a compressor is used for compressing the refrigerant vapor.

REFRIGERATION SYSTEM, SELF-CONTAINED. A complete factory-assembled and tested system that is shipped in one or more sections and has no refrigerant-containing parts that are joined in the field by other than companion or block valves.

REGISTERED DESIGN PROFESSIONAL. An individual who is registered or licensed to practice their respective design profession as defined by the statutory requirements of the professional registration laws of the state or jurisdiction in which the project is to be constructed.

RETURN AIR. Air removed from an approved conditioned space or location and recirculated or exhausted.

RETURN AIR SYSTEM. An assembly of connected ducts, plenums, fittings, registers and grilles through which air from the space or spaces to be heated or cooled is conducted back to the supply unit (see also 'Supply air system').

ROOM HEATER VENTED. A free-standing heating unit burning solid or liquid fuel for direct heating of the space in and adjacent to that in which the unit is located.

SAFETY VALVE. A valve that relieves pressure in a steam boiler by opening fully at the rated discharge pressure. The valve is of the spring-pop type.

SELF-CONTAINED EQUIPMENT. Complete, factory-assembled and tested, heating, air-conditioning or refrigeration equipment installed as a single unit, and having all working parts, complete with motive power, in an enclosed unit of said machinery.

SHAFT. An enclosed space extending through one or more stories of a building, connecting vertical openings in successive floors, or floors and the roof.

SHAFT ENCLOSURE. The walls or construction forming the boundaries of a shaft.

SMOKE DAMPER. A listed device that is designed to resist the passage of air and smoke. The device is arranged to operate automatically, controlled by a smoke detection system, and when required, is capable of being positioned manually from a remote command station.

SMOKE-DEVELOPED INDEX. A numerical value assigned to a material tested in accordance with ASTM E 84.

SOLID FUEL (COOKING APPLICATIONS). Applicable to commercial food service operations only, solid fuel is any bulk material such as hardwood, mesquite, charcoal or briquettes that is combusted to produce heat for cooking operations.

SOURCE CAPTURE SYSTEM. A mechanical exhaust system designed and constructed to capture air contaminants at their source and to exhaust such contaminants to the outdoor atmosphere.

STATIONARY FUEL CELL POWER PLANT. A self-contained package or factory-matched packages which constitute an automatically operated assembly of integrated systems for generating useful electrical energy and recoverable thermal energy that is permanently connected and fixed in place.

STEAM-HEATING BOILER. A boiler operated at pressures not exceeding 15 psi (103 kPa) for steam.

STOP VALVE. A shutoff valve for controlling the flow of liquid or gases.

STORY. That portion of a building included between the upper surface of a floor and the upper surface of the floor next above, except that the topmost story shall be that portion of a building included between the upper surface of the topmost floor and the ceiling or roof above.

STRENGTH, ULTIMATE. The highest stress level that the component will tolerate without rupture.

SUPPLY AIR. That air delivered to each or any space supplied by the air distribution system or the total air delivered to all spaces supplied by the air distribution system, which is provided for ventilating, heating, cooling, humidification, dehumidification and other similar purposes.

SUPPLY AIR SYSTEM. An assembly of connected ducts, plenums, fittings, registers and grilles through which air, heated or cooled, is conducted from the supply unit to the space or spaces to be heated or cooled (see also 'Return air system').

THEORETICAL AIR. The exact amount of air required to supply oxygen for complete combustion of a given quantity of a specific fuel.

THERMAL RESISTANCE (R). A measure of the ability to retard the flow of heat. The R-value is the reciprocal of thermal conductance.

TLV-TWA (THRESHOLD LIMIT VALUE-TIME-WEIGHTED AVERAGE). The time-weighted average concentration of a refrigerant or other chemical in air for a normal 8-hour workday and a 40-hour workweek, to which nearly all workers are repeatedly exposed, day after day, without adverse effects, as adopted by the American Conference of Government Industrial Hygienists (ACGIH).

TOILET ROOM. A room containing a water closet and, frequently, a lavatory, but not a bathtub, shower, spa or similar bathing fixture.

TOXICITY CLASSIFICATION. Refrigerants shall be classified for toxicity to one of two classes in accordance with ASHRAE 34:

Class A. Refrigerants for which toxicity has not been identified at concentrations less than or equal to 400 parts per million (ppm), based on data used to determine Threshold Limit Value-Time-Weighted Average (TLV-TWA) or consistent indices.

Class B. Refrigerants for which there is evidence of toxicity at concentrations below 400 ppm, based on data used to determine TLV-TWA or consistent indices.

TRANSITION FITTINGS, PLASTIC TO STEEL. An adapter for joining plastic pipe to steel pipe. The purpose of this fitting is to provide a permanent, pressure-tight connection between two materials which cannot be joined directly one to another.

UNCONFINED SPACE. A space having a volume not less than 50 cubic feet per 1,000 Btu/h (4.8 m³/kW) of the aggregate input rating of all appliances installed in that space. Rooms communicating directly with the space in which the appliances are installed, through openings not furnished with doors, are considered a part of the unconfined space.

UNIT HEATER. A self-contained appliance of the fan type, designed for the delivery of warm air directly into the space in which the appliance is located.

UNUSUALLY TIGHT CONSTRUCTION. Construction meeting the following requirements:

1. Walls exposed to the outside atmosphere having a continuous water vapor retarder with a rating of 1 perm (57 ng/s • m² • Pa) or less with openings gasketed or sealed;

2. Openable windows and doors meeting the air leakage requirements of the *International Energy Conservation Code*, Section 502.1.4; and

3. Caulking or sealants are applied to areas, such as joints around window and door frames, between sole plates and floors, between wall-ceiling joints, between wall panels, at penetrations for plumbing, electrical and gas lines, and at other openings.

VENT. A pipe or other conduit composed of factory-made components, containing a passageway for conveying combustion products and air to the atmosphere, listed and labeled for use with a specific type or class of appliance.

Pellet vent. A vent listed and labeled for use with listed pellet-fuel-burning appliances.

Type L vent. A vent listed and labeled for use with the following:

1. Oil-burning appliances that are listed for use with Type L vents.

2. Gas-fired appliances that are listed for use with Type B vents.

VENT CONNECTOR. The pipe that connects an approved fuel-fired appliance to a vent.

VENT DAMPER DEVICE, AUTOMATIC. A device intended for installation in the venting system, in the outlet of an individual automatically operated fuel-burning appliance that is designed to open the venting system automatically when the appliance is in operation and to close off the venting system automatically when the appliance is in a standby or shutdown condition.

VENTILATION. The natural or mechanical process of supplying conditioned or unconditioned air to, or removing such air from, any space.

VENTILATION AIR. That portion of supply air that comes from the outside (outdoors), plus any recirculated air that has been treated to maintain the desired quality of air within a designated space.

VENTING SYSTEM. A continuous open passageway from the flue collar of an appliance to the outside atmosphere for the purpose of removing flue or vent gases. A venting system is usually composed of a vent or a chimney and vent connector, if used, assembled to form the open passageway.

WATER HEATER. Any heating appliance or equipment that heats potable water and supplies such water to the potable hot water distribution system.

CHAPTER 3
GENERAL REGULATIONS

SECTION 301
GENERAL

301.1 Scope. This chapter shall govern the approval and installation of all equipment and appliances that comprise parts of the building mechanical systems regulated by this code in accordance with Section 101.2.

301.2 Energy utilization. Heating, ventilating and air-conditioning systems of all structures shall be designed and installed for efficient utilization of energy in accordance with the *International Energy Conservation Code.*

301.3 Fuel gas appliances and equipment. The approval and installation of fuel gas distribution piping and equipment, fuel gas-fired appliances and fuel gas-fired appliance venting systems shall be in accordance with the *International Fuel Gas Code.*

301.4 Listed and labeled. All appliances regulated by this code shall be listed and labeled, unless otherwise approved in accordance with Section 105.

301.5 Labeling. Labeling shall be in accordance with the procedures set forth in Sections 301.5.1 through 301.5.2.3.

301.5.1 Testing. An approved agency shall test a representative sample of the mechanical equipment and appliances being labeled to the relevant standard or standards. The approved agency shall maintain a record of all of the tests performed. The record shall provide sufficient detail to verify compliance with the test standard.

301.5.2 Inspection and identification. The approved agency shall periodically perform an inspection, which shall be in-plant if necessary, of the mechanical equipment and appliances to be labeled. The inspection shall verify that the labeled mechanical equipment and appliances are representative of the mechanical equipment and appliances tested.

301.5.2.1 Independent. The agency to be approved shall be objective and competent. To confirm its objectivity, the agency shall disclose all possible conflicts of interest.

301.5.2.2 Equipment. An approved agency shall have adequate equipment to perform all required tests. The equipment shall be periodically calibrated.

301.5.2.3 Personnel. An approved agency shall employ experienced personnel educated in conducting, supervising and evaluating tests.

301.6 Label information. A permanent factory-applied name-plate(s) shall be affixed to appliances on which shall appear in legible lettering, the manufacturer's name or trademark, the model number, serial number and the seal or mark of the approved agency. A label shall also include the following:

1. Electrical equipment and appliances: Electrical rating in volts, amperes and motor phase; identification of individual electrical components in volts, amperes or watts, motor phase; Btu/h (W) output; and required clearances.

2. Absorption units: Hourly rating in Btu/h (W); minimum hourly rating for units having step or automatic modulating controls; type of fuel; type of refrigerant; cooling capacity in Btu/h (W); and required clearances.

3. Fuel-burning units: Hourly rating in Btu/h (W); type of fuel approved for use with the appliance; and required clearances.

4. Electric comfort heating appliances: Name and trade-mark of the manufacturer; the model number or equivalent; the electric rating in volts, ampacity and phase; Btu/h (W) output rating; individual marking for each electrical component in amperes or watts, volts and phase; required clearances from combustibles; and a seal indicating approval of the appliance by an approved agency.

301.7 Electrical. Electrical wiring, controls and connections to equipment and appliances regulated by this code shall be in accordance with the ICC *Electrical Code.*

301.8 Plumbing connections. Potable water supply and building drainage system connections to equipment and appliances regulated by this code shall be in accordance with the *International Plumbing Code.*

301.9 Fuel types. Fuel-fired appliances shall be designed for use with the type of fuel to which they will be connected and the altitude at which they are installed. Appliances that comprise parts of the building mechanical system shall not be converted for the usage of a different fuel, except where approved and converted in accordance with the manufacturer's instructions. The fuel input rate shall not be increased or decreased beyond the limit rating for the altitude at which the appliance is installed.

301.10 Vibration isolation. Where vibration isolation of equipment and appliances is employed, an approved means of supplemental restraint shall be used to accomplish the support and restraint.

301.11 Repair. Defective material or parts shall be replaced or repaired in such a manner so as to preserve the original approval or listing.

301.12 Wind resistance. Mechanical equipment, appliances and supports that are exposed to wind shall be designed and installed to resist the wind pressures determined in accordance with the *International Building Code.*

[B] 301.13 Flood hazard. For structures located in flood hazard areas, mechanical systems, equipment and appliances shall be located at or above the design flood elevation.

> **Exception:** Mechanical systems, equipment and appliances are permitted to be located below the design flood elevation provided that they are designed and installed to prevent water from entering or accumulating within the components and to resist hydrostatic and hydrodynamic loads and stresses, including the effects of buoyancy, during the oc-

currence of flooding to the design flood elevation in compliance with the flood-resistant construction requirements of the *International Building Code.*

[B] 301.13.1 High-velocity wave action. In flood hazard areas subject to high-velocity wave action, mechanical systems and equipment shall not be mounted on or penetrate walls intended to break away under flood loads.

301.14 Rodent proofing. Buildings or structures and the walls enclosing habitable or occupiable rooms and spaces in which persons live, sleep or work, or in which feed, food or foodstuffs are stored, prepared, processed, served or sold, shall be constructed to protect against the entrance of rodents in accordance with the *International Building Code.*

301.15 Seismic resistance. When earthquake loads are applicable in accordance with the *International Building Code*, mechanical system supports shall be designed and installed for the seismic forces in accordance with the *International Building Code.*

SECTION 302
PROTECTION OF STRUCTURE

302.1 Structural safety. The building or structure shall not be weakened by the installation of mechanical systems. Where floors, walls, ceilings or any other portion of the building or structure are required to be altered or replaced in the process of installing or repairing any system, the building or structure shall be left in a safe structural condition in accordance with the *International Building Code.*

302.2 Penetrations of floor/ceiling assemblies and fire-resistance-rated assemblies. Penetrations of floor/ceiling assemblies and assemblies required to have a fire-resistance rating shall be protected in accordance with the *International Building Code.*

[B] 302.3 Cutting, notching and boring in wood framing. The cutting, notching and boring of wood framing members shall comply with Sections 302.3.1 through 302.3.4.

[B] 302.3.1 Joist notching. Notches on the ends of joists shall not exceed one-fourth the joist depth. Holes bored in joists shall not be within 2 inches (51 mm) of the top or bottom of the joist, and the diameter of any such hole shall not exceed one-third the depth of the joist. Notches in the top or bottom of joists shall not exceed one-sixth the depth and shall not be located in the middle third of the span.

[B] 302.3.2 Stud cutting and notching. In exterior walls and bearing partitions, any wood stud is permitted to be cut or notched not to exceed 25 percent of its depth. Cutting or notching of studs not greater than 40 percent of their depth is permitted in nonbearing partitions supporting no loads other than the weight of the partition.

[B] 302.3.3 Bored holes. A hole not greater in diameter than 40 percent of the stud depth is permitted to be bored in any wood stud. Bored holes not greater than 60 percent of the depth of the stud are permitted in nonbearing partitions or in any wall where each bored stud is doubled, provided not more than two such successive doubled studs are so bored. In no case shall the edge of the bored hole be nearer than 0.625 inch (15.9 mm) to the edge of the stud. Bored holes shall not be located at the same section of stud as a cut or notch.

[B] 302.3.4 Engineered wood products. Cuts, notches and holes bored in trusses, laminated veneer lumber, glue-laminated members and I-joists are prohibited except where the effects of such alterations are specifically considered in the design of the member.

[B] 302.4 Alterations to trusses. Truss members and components shall not be cut, drilled, notched, spliced or otherwise altered in any way without written concurrence and approval of a registered design professional. Alterations resulting in the addition of loads to any member (e.g., HVAC equipment, water heaters) shall not be permitted without verification that the truss is capable of supporting such additional loading.

[B] 302.5 Cutting, notching and boring in steel framing. The cutting, notching and boring of steel framing members shall comply with Sections 302.5.1 through 302.5.3.

[B] 302.5.1 Cutting, notching and boring holes in structural steel framing. The cutting, notching and boring of holes in structural steel framing members shall be as prescribed by the registered design professional.

[B] 302.5.2 Cutting, notching and boring holes in cold-formed steel framing. Flanges and lips of load-bearing cold-formed steel framing members shall not be cut or notched. Holes in webs of load-bearing cold-formed steel framing members shall be permitted along the centerline of the web of the framing member and shall not exceed the dimensional limitations, penetration spacing or minimum hole edge distance as prescribed by the registered design professional. Cutting, notching and boring holes of steel floor/roof decking shall be as prescribed by the registered design professional.

[B] 302.5.3 Cutting, notching and boring holes in non-structural cold-formed steel wall framing. Flanges and lips of nonstructural cold-formed steel wall studs shall not be cut or notched. Holes in webs of nonstructural cold-formed steel wall studs shall be permitted along the center-line of the web of the framing member, shall not exceed 1.5 inches (38 mm) in width or 4 inches (102 mm) in length, and shall not be spaced less than 24 inches (610 mm) center to center from another hole or less than 10 inches (254 mm) from the bearing end.

SECTION 303
EQUIPMENT AND APPLIANCE LOCATION

303.1 General. Equipment and appliances shall be located as required by this section, specific requirements elsewhere in this code and the conditions of the equipment and appliance listing.

303.2 Hazardous locations. Appliances shall not be located in a hazardous location unless listed and approved for the specific installation.

303.3 Prohibited locations. Fuel-fired appliances shall not be located in, or obtain combustion air from, any of the following rooms or spaces:

1. Sleeping rooms.

2. Bathrooms.

3. Toilet rooms.

4. Storage closets.

5. Surgical rooms.

> **Exception:** This section shall not apply to the following appliances:
>
> 1. Direct-vent appliances that obtain all combustion air directly from the outdoors.
>
> 2. Solid fuel-fired appliances, provided that the room is not a confined space and the building is not of unusually tight construction.
>
> 3. Appliances installed in a dedicated enclosure in which all combustion air is taken directly from the outdoors, in accordance with Section 703. Access to such enclosure shall be through a solid door, weather-stripped in accordance with the exterior door air leakage requirements of the *International Energy Conservation Code* and equipped with an approved self-closing device.

303.4 Protection from damage. Appliances shall not be installed in a location where subject to mechanical damage unless protected by approved barriers.

303.5 Indoor locations. Fuel-fired furnaces and boilers installed in closets and alcoves shall be listed for such installation. For purposes of this section, a closet or alcove shall be defined as a room or space having a volume less than 12 times the total volume of fuel-fired appliances other than boilers and less than 16 times the total volume of boilers. Room volume shall be computed using the gross floor area and the actual ceiling height up to a maximum computation height of 8 feet (2438 mm).

303.6 Outdoor locations. Appliances installed in other than indoor locations shall be listed and labeled for outdoor installation.

303.7 Pit locations. Appliances installed in pits or excavations shall not come in direct contact with the surrounding soil. The sides of the pit or excavation shall be held back a minimum of 12 inches (305 mm) from the appliance. Where the depth exceeds 12 inches (305 mm) below adjoining grade, the walls of the pit or excavation shall be lined with concrete or masonry. Such concrete or masonry shall extend a minimum of 4 inches (102 mm) above adjoining grade and shall have sufficient lateral load-bearing capacity to resist collapse. The appliance shall be protected from flooding in an approved manner.

[B] 303.8 Elevator shafts. Mechanical systems shall not be located in an elevator shaft.

SECTION 304
INSTALLATION

304.1 General. Equipment and appliances shall be installed as required by the terms of their approval, in accordance with the conditions of the listing, the manufacturer's installation instructions and this code. Manufacturer's installation instructions shall be available on the job site at the time of inspection.

304.2 Conflicts. Where conflicts between this code and the conditions of listing or the manufacturer's installation instructions occur, the provisions of this code shall apply.

> **Exception:** Where a code provision is less restrictive than the conditions of the listing of the equipment or appliance or the manufacturer's installation instructions, the conditions of the listing and the manufacturer's installation instructions shall apply.

304.3 Elevation of ignition source. Equipment and appliances having an ignition source and located in hazardous locations and public garages, private garages, repair garages, automotive motor-fuel-dispensing facilities and parking garages shall be elevated such that the source of ignition is not less than 18 inches (457 mm) above the floor surface on which the equipment or appliance rests. Such equipment and appliances shall not be installed in Group H occupancies or control areas where open use, handling or dispensing of combustible, flammable or explosive materials occurs. For the purpose of this section, rooms or spaces that are not part of the living space of a dwelling unit and that communicate directly with a private garage through openings shall be considered to be part of the private garage.

304.4 Hydrogen generating and refueling operations. Ventilation shall be required in accordance with Section 304.4.1, 304.4.2 or 304.4.3 in public garages, private garages, repair garages, automotive motor-fuel-dispensing facilities and parking garages which contain hydrogen generating appliances or refueling systems. Such spaces shall be used for the storage of not more than three hydrogen-fueled passenger motor vehicles and have a floor area not exceeding 850 square feet (79 square meters). The maximum rated output capacity of hydrogen generating appliances shall not exceed 4 SCFM of hydrogen for each 250 square feet (23.2 square meters) of floor area in such spaces. Such equipment and appliances shall not be installed in Group H occupancies except where the occupancy is specifically designed for hydrogen use, or in control areas where open-use, handling or dispensing of combustible, flammable or explosive materials occurs. For the purpose of this section, rooms or spaces that are not part of the living space of a dwelling unit and that communicate directly with a private garage through openings shall be considered to be part of the private garage.

304.4.1 Natural ventilation. Indoor locations intended for hydrogen generating or refueling operations shall communicate with the outdoors in accordance with Sections 304.4.1.1 and 304.4.1.2. The minimum dimension of air openings shall be not less than 3 inches (76 mm). Where ducts are used, they shall be of the same cross-sectional area as the free area of the openings to which they connect. In such locations, equipment and appliances having an ignition source shall be located such that the source of ignition is not less than 12 inches (228 mm) below the ceiling.

304.4.1.1 Two openings. Two permanent openings, one located entirely within 12 inches (305 mm) of the ceiling of the garage, and one located entirely within 12 inches (305 mm) of the floor of the garage, shall be provided in the same exterior wall. The openings shall communicate directly, or by ducts, with the outdoors. Each opening shall directly communicate with the outdoors horizontally, and have a minimum free area of $^1/_2$ square foot per 1,000 cubic feet (1641 mm^2/m^3) of garage volume.

304.4.1.2 Louvers and grilles. In calculating free area required by Section 304.4.1, the required size of openings shall be based on the net free area of each opening. If the free area through a design of louver or grille is known, it shall be used in calculating the size opening required to provide the free area specified. If the design and free area are not known, it shall be assumed that wood louvers will have 25 percent free area and metal louvers and grilles will have 75 percent free area. Louvers and grilles shall be fixed in the open position.

304.4.2 Mechanical ventilation. Indoor locations intended for hydrogen generating or refueling operations shall be ventilated in accordance with Section 502.16.

304.4.3 Specially engineered installations. As an alternative to the provisions of Sections 304.4.1 and 304.4.2 the necessary supply of air for ventilation and dilution of flammable gases shall be provided by an approved engineered system.

304.5 Public garages. Appliances located in public garages, motor fuel-dispensing facilities, repair garages or other areas frequented by motor vehicles, shall be installed a minimum of 8 feet (2438 mm) above the floor. Where motor vehicles exceed 6 feet (1829 mm) in height and are capable of passing under an appliance, appliances shall be installed a minimum of 2 feet (610 mm) higher above the floor than the height of the tallest vehicle.

Exception: The requirements of this section shall not apply where the appliances are protected from motor vehicle impact and installed in accordance with Section 304.3 and NFPA 88B.

304.6 Private garages. Appliances located in private garages and carports shall be installed with a minimum clearance of 6 feet (1829 mm) above the floor.

Exception: The requirements of this section shall not apply where the appliances are protected from motor vehicle impact and installed in accordance with Section 304.3.

304.7 Construction and protection. Boiler rooms and furnace rooms shall be protected as required by the *International Building Code*.

304.8 Clearances to combustible construction. Heat-producing equipment and appliances shall be installed to maintain the required clearances to combustible construction as specified in the listing and manufacturer's instructions. Such clearances shall be reduced only in accordance with Section 308. Clearances to combustibles shall include such considerations as door swing, drawer pull, overhead projections or shelving and window swing, shutters, coverings and drapes. Devices such as doorstops or limits, closers, drapery ties or guards shall not be used to provide the required clearances.

304.9 Clearances from grade. Equipment and appliances installed at grade level shall be supported on a level concrete slab or other approved material extending above adjoining grade or shall be suspended a minimum of 6 inches (152 mm) above adjoining grade.

304.10 Guards. Guards shall be provided where appliances, equipment, fans or other components that require service are located within 10 feet (3048 mm) of a roof edge or open side of a walking surface and such edge or open side is located more than 30 inches (762 mm) above the floor, roof or grade below. The guard shall extend not less than 30 inches (762 mm) beyond each end of such appliance, equipment, fan or component and the top of the guard shall be located not less than 42 inches (1067 mm) above the elevated surface adjacent to the guard. The guard shall be constructed so as to prevent the passage of a 21-inch-diameter (533 mm) sphere and shall comply with the loading requirements for guards specified in the *International Building Code*.

304.11 Area served. Appliances serving different areas of a building other than where they are installed shall be permanently marked in an approved manner that uniquely identifies the appliance and the area it serves.

SECTION 305
PIPING SUPPORT

305.1 General. All mechanical system piping shall be supported in accordance with this section.

305.2 Materials. Pipe hangers and supports shall have sufficient strength to withstand all anticipated static and specified dynamic loading conditions associated with the intended use. Pipe hangers and supports that are in direct contact with piping shall be of approved materials that are compatible with the piping and that will not promote galvanic action.

305.3 Structural attachment. Hangers and anchors shall be attached to the building construction in an approved manner.

305.4 Interval of support. Piping shall be supported at distances not exceeding the spacing specified in Table 305.4, or in accordance with MSS SP-69.

305.5 Protection against physical damage. In concealed locations where piping, other than cast-iron or steel, is installed through holes or notches in studs, joists, rafters or similar members less than 1.5 inches (38 mm) from the nearest edge of the member, the pipe shall be protected by shield plates. Protective shield plates shall be a minimum of 0.062-inch-thick (1.6 mm) steel, shall cover the area of the pipe where the member is notched or bored, and shall extend a minimum of 2 inches (51 mm) above sole plates and below top plates.

SECTION 306
ACCESS AND SERVICE SPACE

306.1 Clearances for maintenance and replacement. Clearances around appliances to elements of permanent construction, including other installed equipment and appliances, shall be sufficient to allow inspection, service, repair or replacement without removing such elements of permanent construction or disabling the function of a required fire-resistance-rated assembly.

TABLE 305.4
PIPING SUPPORT SPACING[a]

PIPING MATERIAL	MAXIMUM HIORIZONTAL SPACING (feet)	MAXIMUM VERTICAL SPACING (feet)
ABS pipe	4	10[c]
Aluminum pipe and tubing	10	15
Brass pipe	10	10
Brass tubing, 1-$\frac{1}{4}$-inch diameter and smaller	6	10
Brass tubing, 1$\frac{1}{2}$-inch diameter and larger	10	10
Cast-iron pipe[b]	5	15
Copper or copper-alloy pipe	12	10
Copper or copper-alloy tubing, 1$\frac{1}{4}$-inch diameter and smaller	6	10
Copper or copper-alloy tubing, 1$\frac{1}{2}$-inch diameter and larger	10	10
CPVC pipe or tubing, 1 inch and smaller	3	10[c]
CPVC pipe or tubing 1$\frac{1}{4}$-inch and larger	4	10[c]
Steel pipe	12	15
Steel tubing	8	10
Lead pipe	Continuous	4
PB pipe or tubing	$2\frac{2}{3}$ (32 inches)	4
PEX tubing	$2\frac{2}{3}$ (32 inches)	10[c]
PVC pipe	4	10[c]

For SI: 1 inch = 25.4 mm, 1 foot = 304.8 mm.

a. See Section 301.14.

b. The maximum horizontal spacing of cast-iron pipe hangers shall be increased to 10 feet where 10-foot lengths of pipe are installed.

c. Mid-story guide.

306.1.1 Central furnaces. Central furnaces within compartments or alcoves shall have a minimum working space clearance of 3 inches (76 mm) along the sides, back and top with a total width of the enclosing space being at least 12 inches (305 mm) wider than the furnace. Furnaces having a firebox open to the atmosphere shall have at least 6 inches (152 mm) working space along the front combustion chamber side. Combustion air openings at the rear or side of the compartment shall comply with the requirements of Chapter 7.

Exception: This section shall not apply to replacement appliances installed in existing compartments and alcoves where the working space clearances are in accordance with the equipment or appliance manufacturer's installation instructions.

306.2 Appliances in rooms. Rooms containing appliances requiring access shall be provided with a door and an unobstructed

passageway measuring not less than 36 inches (914 mm) wide and 80 inches (2032 mm) high.

Exception: Within a dwelling unit, appliances installed in a compartment, alcove, basement or similar space shall be accessed by an opening or door and an unobstructed passageway measuring not less than 24 inches (610 mm) wide and large enough to allow removal of the largest appliance in the space, provided that a level service space of not less than 30 inches (762 mm) deep and the height of the appliance, but not less than 30 inches (762 mm), is present at the front or service side of the appliance with the door open.

306.3 Appliances in attics. Attics containing appliances requiring access shall be provided with an opening and unobstructed passageway large enough to allow removal of the largest appliance. The passageway shall not be less than 30 inches (762 mm) high and 22 inches (559 mm) wide and not more than 20 feet (6096 mm) in length measured along the centerline of the passageway from the opening to the appliance. The passageway shall have continuous solid flooring not less than 24 inches (610 mm) wide. A level service space not less than 30 inches (762 mm) deep and 30 inches (762 mm) wide shall be present at the front or service side of the appliance. The clear access opening dimensions shall be a minimum of 20 inches by 30 inches (508 mm by 762 mm), where such dimensions are large enough to allow removal of the largest appliance.

Exception: The passageway and level service space are not required where the appliance is capable of being serviced and removed through the required opening.

306.3.1 Electrical requirements. A lighting fixture controlled by a switch located at the required passageway opening and a receptacle outlet shall be provided at or near the appliance location in accordance with the ICC *Electrical Code.*

306.4 Appliances under floors. Underfloor spaces containing appliances requiring access shall be provided with an access opening and unobstructed passageway large enough to remove the largest appliance. The passageway shall not be less than 30 inches (762 mm) high and 22 inches (559 mm) wide, nor more than 20 feet (6096 mm) in length measured along the centerline of the passageway from the opening to the appliance. A level service space not less than 30 inches (762 mm) deep and 30 inches (762 mm) wide shall be present at the front or service side of the appliance. If the depth of the passageway or the service space exceeds 12 inches (305 mm) below the adjoining grade, the walls of the passageway shall be lined with concrete or masonry. Such concrete or masonry shall extend a minimum of 4 inches (102 mm) above the adjoining grade and shall have sufficient lateral-bearing capacity to resist collapse. The clear access opening dimensions shall be a minimum of 22 inches by 30 inches (559 mm by 762 mm), where such dimensions are large enough to allow removal of the largest appliance.

Exception: The passageway is not required where the level service space is present when the access is open and the appliance is capable of being serviced and removed through the required opening.

306.4.1 Electrical requirements. A lighting fixture controlled by a switch located at the required passageway opening and a receptacle outlet shall be provided at or near the appliance location in accordance with the ICC *Electrical Code.*

306.5 Equipment and appliances on roofs or elevated structures. Where equipment and appliances requiring access are installed on roofs or elevated structures at a height exceeding 16 feet (4877 mm), such access shall be provided by a permanent approved means of access, the extent of which shall be from grade or floor level to the equipment and appliances' level service space. Such access shall not require climbing over obstructions greater than 30 inches (762 mm) high or walking on roofs having a slope greater than 4 units vertical in 12 units horizontal (33-percent slope).

Permanent ladders installed to provide the required access shall comply with the following minimum design criteria:

1. The side railing shall extend above the parapet or roof edge not less than 30 inches (762 mm).

2. Ladders shall have rung spacing not to exceed 14 inches (356 mm) on center.

3. Ladders shall have a toe spacing not less than 6 inches (152 mm) deep.

4. There shall be a minimum of 18 inches (457 mm) between rails.

5. Rungs shall have a minimum 0.75-inch (19 mm) diameter and be capable of withstanding a 300-pound (136.1 kg) load.

6. Ladders over 30 feet (9144 mm) in height shall be provided with offset sections and landings capable of withstanding 100 pounds (488.2 kg/m²) per square foot.

7. Ladders shall be protected against corrosion by approved means.

Catwalks installed to provide the required access shall be not less than 24 inches (610 mm) wide and shall have railings as required for service platforms.

Exception: This section shall not apply to Group R-3 occupancies.

306.6 Sloped roofs. Where appliances are installed on a roof having a slope of 3 units vertical in 12 units horizontal (25-percent slope) or greater and having an edge more than 30 inches (762 mm) above grade at such edge, a level platform shall be provided on each side of the appliance to which access is required by the manufacturer's installation instructions for service, repair or maintenance. The platform shall not be less than 30 inches (762 mm) in any dimension and shall be provided with guards in accordance with Section 304.10.

SECTION 307
CONDENSATE DISPOSAL

307.1 Fuel-burning appliances. Liquid combustion by-products of condensing appliances shall be collected and discharged to an approved plumbing fixture or disposal area in accordance with the manufacturer's installation instructions. Condensate piping shall be of approved corrosion-resistant material and shall not be smaller than the drain connection on the appliance. Such piping shall maintain a minimum horizontal slope in the direction of discharge of not less than one-eighth unit vertical in 12 units horizontal (1-percent slope).

307.2 Evaporators and cooling coils. Condensate drain systems shall be provided for equipment and appliances containing evaporators or cooling coils. Condensate drain systems shall be designed, constructed and installed in accordance with Sections 307.2.1 through 307.2.4.

307.2.1 Condensate disposal. Condensate from all cooling coils and evaporators shall be conveyed from the drain pan outlet to an approved place of disposal. Condensate shall not discharge into a street, alley or other areas so as to cause a nuisance.

307.2.2 Drain pipe materials and sizes. Components of the condensate disposal system shall be cast iron, galvanized steel, copper, cross-linked polyethylene, polybutylene, polyethylene, ABS, CPVC or PVC pipe or tubing. All components shall be selected for the pressure and temperature rating of the installation. Condensate waste and drain line size shall be not less than $^3/_4$-inch (19 mm) internal diameter and shall not decrease in size from the drain pan connection to the place of condensate disposal. Where the drain pipes from more than one unit are manifolded together for condensate drainage, the pipe or tubing shall be sized in accordance with an approved method. All horizontal sections of drain piping shall be installed in uniform alignment at a uniform slope.

307.2.3 Auxiliary and secondary drain systems. In addition to the requirements of Section 307.2.1, a secondary drain or auxiliary drain pan shall be required for each cooling or evaporator coil where damage to any building components will occur as a result of overflow from the equipment drain pan or stoppage in the condensate drain piping. One of the following methods shall be used:

1. An auxiliary drain pan with a separate drain shall be provided under the coils on which condensation will occur. The auxiliary pan drain shall discharge to a conspicuous point of disposal to alert occupants in the event of a stoppage of the primary drain. The pan shall have a minimum depth of 1.5 inches (38 mm), shall not be less than 3 inches (76 mm) larger than the unit or the coil dimensions in width and length and shall be constructed of corrosion-resistant material. Metallic pans shall have a minimum thickness of not less than 0.0276-inch (0.7 mm) galvanized sheet metal. Nonmetallic pans shall have a minimum thickness of not less than 0.0625 inch (1.6 mm).

2. A separate overflow drain line shall be connected to the drain pan provided with the equipment. Such overflow drain shall discharge to a conspicuous point of disposal to alert occupants in the event of a stoppage of the primary drain. The overflow drain line shall connect to the drain pan at a higher level than the primary drain connection.

3. An auxiliary drain pan without a separate drain line shall be provided under the coils on which condensate will occur. Such pan shall be equipped with a water-level detection device that will shut off the equipment served prior to overflow of the pan. The auxiliary drain pan shall be constructed in accordance with Item 1 of this section.

307.2.4 Traps. Condensate drains shall be trapped as required by the equipment or appliance manufacturer.

SECTION 308
CLEARANCE REDUCTION

308.1 Scope. This section shall govern the reduction in required clearances to combustible materials and combustible assemblies for chimneys, vents, kitchen exhaust equipment, mechanical appliances, and mechanical devices and equipment.

308.2 Listed appliances and equipment. The reduction of the required clearances to combustibles for listed and labeled appliances and equipment shall be in accordance with the requirements of this section except that such clearances shall not be reduced where reduction is specifically prohibited by the terms of the appliance or equipment listing.

308.3 Protective assembly construction and installation. Reduced clearance protective assemblies, including structural and support elements, shall be constructed of noncombustible materials. Spacers utilized to maintain an airspace between the protective assembly and the protected material or assembly shall be noncombustible. Where a space between the protective assembly and protected combustible material or assembly is specified, the same space shall be provided around the edges of the protective assembly and the spacers shall be placed so as to allow air circulation by convection in such space. Protective assemblies shall not be placed less than 1 inch (25 mm) from the mechanical appliances, devices or equipment, regardless of the allowable reduced clearance.

308.4 Allowable reduction. The reduction of required clearances to combustible assemblies or combustible materials shall be based on the utilization of a reduced clearance protective assembly in accordance with Section 308.5 or 308.6.

308.5 Labeled assemblies. The allowable clearance reduction shall be based on an approved reduced clearance protective assembly that has been tested and bears the label of an approved agency.

308.6 Reduction table. The allowable clearance reduction shall be based on one of the methods specified in Table 308.6. Where required clearances are not listed in Table 308.6, the reduced clearances shall be determined by linear interpolation between the distances listed in the table. Reduced clearances shall not be derived by extrapolation below the range of the table.

308.7 Solid fuel-burning appliances. The clearance reduction methods specified in Table 308.6 shall not be utilized to reduce the clearance required for solid fuel-burning appliances that are labeled for installation with clearances of 12 inches (305 mm) or less. Where appliances are labeled for installation with clearances of greater than 12 inches (305 mm), the clearance reduction methods of Table 308.6 shall not reduce the clearance to less than 12 inches (305 mm).

308.8 Masonry chimneys. The clearance reduction methods specified in Table 308.6 shall not be utilized to reduce the clearances required for masonry chimneys as specified in Chapter 8 and the *International Building Code*.

308.9 Chimney connector pass-throughs. The clearance reduction methods specified in Table 308.6 shall not be utilized to reduce the clearances required for chimney connector pass-throughs as specified in Section 803.10.4.

308.10 Masonry fireplaces. The clearance reduction methods specified in Table 308.6 shall not be utilized to reduce the clearances required for masonry fireplaces as specified in Chapter 8 and the *International Building Code*.

308.11 Kitchen exhaust ducts. The clearance reduction methods specified in Table 308.6 shall not be utilized to reduce the minimum clearances required by Section 506.3.10 for kitchen exhaust ducts enclosed in a shaft.

[B] SECTION 309
TEMPERATURE CONTROL

[B] 309.1 Space-heating systems. Interior spaces intended for human occupancy shall be provided with active or passive space-heating systems capable of maintaining a minimum indoor temperature of 68°F (20°C) at a point 3 feet (914 mm) above floor on the design heating day. The installation of portable space heaters shall not be used to achieve compliance with this section.

Exception: Interior spaces where the primary purpose is not associated with human comfort.

[F] SECTION 310
EXPLOSION CONTROL

[F] 310.1 Required. Structures occupied for purposes involving explosion hazards shall be provided with explosion control where required by the *International Fire Code*. Explosion control systems shall be designed and installed in accordance with the *International Fire Code*.

[F] SECTION 311
SMOKE AND HEAT VENTS

[F] 311.1 Required. Approved smoke and heat vents shall be installed in the roofs of one-story buildings where required by the *International Fire Code*. Smoke and heat vents shall be designed and installed in accordance with the *International Fire Code*.

SECTION 312
HEATING AND COOLING LOAD CALCULATIONS

312.1 Load calculations. Heating and cooling system design loads for the purpose of sizing systems, appliances and equipment shall be determined in accordance with the procedures described in the ASHRAE *Handbook of Fundamentals*. Heating and cooling loads shall be adjusted to account for load reductions that are achieved when energy recovery systems are utilized in the HVAC system in accordance with the ASHRAE Handbook - *HVAC Systems and Equipment*. Alternatively, design loads shall be determined by an approvedequivalent computation procedure, using the design parameters specified in Chapter 3 of the *International Energy Conservation Code*.

TABLE 308.6
CLEARANCE REDUCTION METHODS

TYPE OF PROTECTIVE ASSEMBLY[a]	REDUCED CLEARANCE WITH PROTECTION (inches)[a]							
	Horizontal combustible assemblies located above the heat source				Horizontal combustible assemblies located beneath the heat source and all vertical combustible assemblies			
	Required clearance to combustibles without protection (inches)[a]				Required clearance to combustible without protection (inches)[a]			
	36	18	9	6	36	18	9	6
Galvanized sheet metal, minimum nominal thickness of 0.024 inch (No. 24 Gage), mounted on 1-inch glass fiber or mineral wool batt reinforced with wire on the back, 1 inch off the combustible assembly	18	9	5	3	12	6	3	3
Galvanized sheet metal, minimum nominal thickness of 0.024 inch (No. 24 Gage), spaced 1 inch off the combustible assembly	18	9	5	3	12	6	3	2
Two layers of galvanized sheet metal, minimum nominal thickness of 0.024 inch (No. 24 Gage), having a 1-inch airspace between layers, spaced 1 inch off the combustible assembly	18	9	5	3	12	6	3	3
Two layers of galvanized sheet metal, minimum nominal thickness of 0.024 inch (No. 24 Gage), having 1 inch of fiberglass insulation between layers, spaced 1 inch off the combustible assembly	18	9	5	3	12	6	3	3
0.5-inch inorganic insulating board, over 1 inch of fiberglass or mineral wool batt, against the combustible assembly	24	12	6	4	18	9	5	3
3.5-inch brick wall, spaced 1 inch off the combustible wall	—	—	—	—	12	6	6	6
3.5-inch brick wall, against the combustible wall	—	—	—	—	24	12	6	5

For SI: 1 inch = 25.4 mm, °C = [(°F)-32]/1.8, 1 pound per cubic foot = 16.02 kg/m^3, 1.0 Btu • in/ft^2 • h • °F = 0.144 W/m^2 • K.

a.   Mineral wool and glass fiber batts (blanket or board) shall have a minimum density of 8 pounds per cubic foot and a minimum melting point of 1,500°F. Insulation material utilized as part of a clearance reduction system shall have a thermal conductivity of 1.0 Btu • in./(ft^2 • h • °F) or less. Insulation board shall be formed of noncombustible material.

CHAPTER 4

VENTILATION

SECTION 401
GENERAL

401.1 Scope. This chapter shall govern the ventilation of spaces within a building intended to be occupied. This chapter does not govern the requirements for smoke control systems.

401.2 Ventilation required. Every occupied space shall be ventilated by natural means in accordance with Section 402 or by mechanical means in accordance with Section 403.

401.3 When required. Ventilation shall be provided during the periods that the room or space is occupied.

[B] 401.4 Exits. Equipment and ductwork for exit enclosure ventilation shall comply with one of the following items:

1. Such equipment and ductwork shall be located exterior to the building and shall be directly connected to the exit enclosure by ductwork enclosed in construction as required by the *International Building Code* for shafts.

2. Where such equipment and ductwork is located within the exit enclosure, the intake air shall be taken directly from the outdoors and the exhaust air shall be discharged directly to the outdoors, or such air shall be conveyed through ducts enclosed in construction as required by the *International Building Code* for shafts.

3. Where located within the building, such equipment and ductwork shall be separated from the remainder of the building, including other mechanical equipment, with construction as required by the *International Building Code* for shafts.

In each case, openings into fire-resistance-rated construction shall be limited to those needed for maintenance and operation and shall be protected by self-closing fire-resistance-rated devices in accordance with the *International Building Code* for enclosure wall opening protectives.

Exit enclosure ventilation systems shall be independent of other building ventilation systems.

401.5 Opening location. Outdoor air exhaust and intake openings shall be located a minimum of 10 feet (3048 mm) from lot lines or buildings on the same lot. Where openings front on a street or public way, the distance shall be measured to the centerline of the street or public way.

Exception: Group R-3.

401.5.1 Intake openings. Mechanical and gravity outdoor air intake openings, shall be located a minimum of 10 feet (3048 mm) from any hazardous or noxious contaminant such as vents, chimneys, plumbing vents, streets, alleys, parking lots and loading docks, except as otherwise specified in this code. Where a source of contaminant is located within 10 feet (3048 mm) of an intake opening, such opening shall be located a minimum of 2 feet (610 mm) below the contaminant source.

401.5.2 Exhaust openings. Outdoor exhaust openings shall be located so as not to create a nuisance. Exhaust air shall not be directed onto walkways.

[B] 401.5.3 Flood hazard. For structures located in flood hazard areas, outdoor exhaust openings shall be at or above the design flood elevation.

401.6 Outdoor opening protection. Air exhaust and intake openings that terminate outdoors shall be protected with corrosion-resistant screens, louvers or grilles. Openings in louvers, grilles and screens shall be sized in accordance with Table 401.6, and shall be protected against local weather conditions. Outdoor air exhaust and intake openings located in exterior walls shall meet the provisions for exterior wall opening protectives in accordance with the *International Building Code*.

TABLE 401.6
OPENING SIZES IN LOUVERS, GRILLES AND SCREENS PROTECTING OUTDOOR EXHAUST AND AIR INTAKE OPENINGS

OUTDOOR OPENING TYPE	MINIMUM AND MAXIMUM OPENING SIZES IN LOUVERS, GRILLES AND SCREENS MEASURED IN ANY DIRECTION
Exhaust openings	Not $< \frac{1}{4}$ inch and not $> \frac{1}{2}$ inch
Intake openings in residential occpancies	Not $< \frac{1}{4}$ inch and not $> \frac{1}{2}$ inch
Intake openings in other than residential occupancies	$> \frac{1}{4}$ inch and not > 1 inch

For SI: 1 inch = 25.4 mm.

401.7 Contaminant sources. Stationary local sources producing air-borne particulates, heat, odors, fumes, spray, vapors, smoke or gases in such quantities as to be irritating or injurious to health shall be provided with an exhaust system in accordance with Chapter 5 or a means of collection and removal of the contaminants. Such exhaust shall discharge directly to an approved location at the exterior of the building.

[B] SECTION 402
NATURAL VENTILATION

402.1 Natural ventilation. Natural ventilation of an occupied space shall be through windows, doors, louvers or other openings to the outdoors. The operating mechanism for such openings shall be provided with ready access so that the openings are readily controllable by the building occupants.

402.2 Ventilation area required. The minimum openable area to the outdoors shall be 4 percent of the floor area being ventilated.

402.3 Adjoining spaces. Where rooms and spaces without openings to the outdoors are ventilated through an adjoining room, the opening to the adjoining rooms shall be unobstructed and shall have an area not less than 8 percent of the floor area of the interior room or space, but not less than 25 square feet (2.3 m²). The minimum openable area to the outdoors shall be based on the total floor area being ventilated.

402.4 Openings below grade. Where openings below grade provide required natural ventilation, the outside horizontal clear space measured perpendicular to the opening shall be one and one-half times the depth of the opening. The depth of the opening shall be measured from the average adjoining ground level to the bottom of the opening.

SECTION 403
MECHANICAL VENTILATION

403.1 Ventilation system. Mechanical ventilation shall be provided by a method of supply air and return or exhaust air. The amount of supply air shall be approximately equal to the amount of return and exhaust air. The system shall not be prohibited from producing negative or positive pressure. The system to convey ventilation air shall be designed and installed in accordance with Chapter 6.

Ventilation supply systems shall be designed to deliver the required rate of supply air to the occupied zone within an occupied space. The occupied zone shall have boundaries measured at 3 inches (76 mm) and 72 inches (1829 mm) above the floor and 24 inches (610 mm) from the enclosing walls.

403.2 Outdoor air required. The minimum ventilation rate of required outdoor air shall be determined in accordance with Section 403.3.

403.2.1 Recirculation of air. The air required by Section 403.3 shall not be recirculated. Air in excess of that required by Section 403.3 shall not be prohibited from being recirculated as a component of supply air to building spaces, except that:

1. Ventilation air shall not be recirculated from one dwelling unit to another or to dissimilar occupancies.

2. Supply air to a swimming pool and associated deck areas shall not be recirculated unless such air is dehumidified to maintain the relative humidity of the area at 60 percent or less. Air from this area shall not be recirculated to other spaces.

3. Where mechanical exhaust is required by Table 403.3, recirculation of air from such spaces shall be prohibited. All air supplied to such spaces shall be exhausted, including any air in excess of that required by Table 403.3.

403.2.2 Transfer air. Except where recirculation from such spaces is prohibited by Table 403.3, air transferred from occupied spaces is not prohibited from serving as makeup air for required exhaust systems in such spaces as kitchens, baths, toilet rooms, elevators and smoking lounges. The amount of transfer air and exhaust air shall be sufficient to provide the flow rates as specified in Sections 403.3 and 403.3.1. The required outdoor air rates specified in Table 403.3 shall be introduced directly into such spaces or into the occupied spaces from which air is transferred or a combination of both.

403.3 Ventilation rate. Ventilation systems shall be designed to have the capacity to supply the minimum outdoor airflow rate determined in accordance with Table 403.3 based on the occupancy of the space and the occupant load or other parameter as stated therein. The occupant load utilized for design of the ventilation system shall not be less than the number determined from the estimated maximum occupant load rate indicated in Table 403.3. Ventilation rates for occupancies not represented in Table 403.3 shall be determined by an approved engineering analysis. The ventilation system shall be designed to supply the required rate of ventilation air continuously during the period the building is occupied, except as otherwise stated in other provisions of the code.

Exception: The occupant load is not required to be determined, based on the estimated maximum occupant load rate indicated in Table 403.3 where approved statistical data document the accuracy of an alternate anticipated occupant density.

403.3.1 System operation. The minimum flow rate of outdoor air that the ventilation system must be capable of supplying during its operation shall be permitted to be based on the rate per person indicated in Table 403.3 and the actual number of occupants present.

403.3.2 Common ventilation system. Where spaces having different ventilation rate requirements are served by a common ventilation system, the ratio of outdoor air to total supply air for the system shall be determined based on the space having the largest outdoor air requirement or shall be determined in accordance with the following formula:

$$Y = \frac{X}{(1 + X - Z)} \qquad \text{(Equation 4-1)}$$

where

$Y = V_{ot}/V_{st}$ = Corrected fraction of outdoor air in system supply.

$X = V_{on}/V_{st}$ = Uncorrected fraction of outdoor air in system supply

$Z = V_{oc}/V_{sc}$ = Fraction of outdoor air in critical space. The critical space is that space with the greatest required fraction of outdoor air in the supply to this space.

V_{ot} = Corrected total outdoor airflow rate.

V_{st} = Total supply flow rate, i.e., the sum of all supply for all branches of the system.

V_{on} = Sum of outdoor airflow rates for all branches on system.

V_{oc} = Outdoor airflow rate required in critical spaces.

V_{sc} = Supply flow rate in critical space.

TABLE 403.3
REQUIRED OUTDOOR VENTILATION AIR

OCCUPANCY CLASSIFICATION	ESTIMATED MAXIMUM OCCUPANT LOAD, PERSONS PER 1,000 SQUARE FEET[a]	OUTDOOR AIR (Cubic feet per Minute (cfm) Per person) UNLESS NOTED[e]
Correctional facilities		
Cells		
without plumbing fixtures	20	20
with plumbing fixtures	20	20
Dining halls	100	15
Guard stations	40	15
Dry Cleaners, laundries		
Coin-operated dry cleaner	20	15
Coin-operated laundries	20	15
Commercial dry cleaner	30	30
Commercial laundry	10	25
Storage, pick up	30	35
Education		
Auditoriums	150	15
Classrooms	50	15
Corridors	—	0.10 cfm/ft^2
Laboratories	30	20
Libraries	20	15
Locker rooms[b]	—	0.50 cfm/ft^2
Music rooms	50	15
Smoking lounges[b,g]	70	60
Training shops	30	20
Food and beverage service		
Bars, cocktail lounges	100	30
Cafeteria, fast food	100	20
Dining rooms	70	20
Kitchens (cooking)[f,g]	20	15
Hospitals, nursing and convalescent homes		
Autopsy rooms[b]	—	0.50 cfm/ft^2
Medical procedure rooms	20	15
Operating rooms	20	30
Patient rooms	10	25
Physical therapy	20	15
Recovery and ICU	20	15
Hotels, motels, resorts and dormitories		
Assembly rooms	120	15
Bathrooms[b,g]	—	35 cfm per room
Bedrooms	—	30 cfm per room
Conference rooms	50	20
Dormitory sleeping areas	20	15
Gambling casinos	120	30
Living rooms	—	30 cfm per room
Lobbies	30	15
Offices		
Conference rooms	50	20
Office spaces	7	20
Reception areas	60	15
Telecommunication centers and data entry	60	20

(continued)

TABLE 403.3—continued
REQUIRED OUTDOOR VENTILATION AIR

OCCUPANCY CLASSIFICATION	ESTIMATED MAXIMUM OCCUPANT LOAD, PERSONS PER 1,000 SQUARE FEET[a]	OUTDOOR AIR (Cubic feet per Minute (cfm) Per person) UNLESS NOTED[e]
Private dwellings, single and Multiple		
Living areas[c]	Based upon number of bedrooms. first bedroom: 2; each additional bedroom: 1	0.35 air changes per hour[a] or 15 cfm per person, whichever is greater
Kitchens[g]	—	100 cfm intermittent or 25 cfm continuous
Toilet rooms and bathrooms[g]	—	mechanical exhaust capacity of 50 cfm interittent or 20 cfm continuous
Garages, separate for each dwelling	—	100 cfm per car
Garages, common for multiple units[b]	—	1.5 cfm/ft^2
Public spaces		
Corridors and utilities	—	0.05 cfm/ft^2
Elevators[g]	—	1.00 cfm/ft^2
Locker rooms[b]	—	0.5 cfm/ft^2
Toilet rooms[b,g]	—	75 cfm per water closet or urinal
Shower room (per shower head)[b,g]	—	50 cfm intermediate or 20 cfm continuous
Smoking lounges[b,g]	70	60
Retail stores, sales floors and Showroom floors		
Basement and street	—	0.30 cfm/ft^2
Dressing rooms	—	0.20 cfm/ft^2
Malls and arcades	—	0.20 cfm/ft^2
Shipping and receiving	—	0.15 cfm/ft^2
Smoking lounges[b]	70	60
Storage rooms	—	0.15 cfm/ft^2
Upper floors	—	0.20 cfm/ft^2
Warehouses	—	0.05 cfm/ft^2
Specialty shops		
Automotive service stations	—	1.5 cfm/ft^2
Barber	25	15
Beauty	25	25
Clothiers, furniture	—	0.30 cfm/ft^2
Florists	8	15
Hardware, drugs, fabrics	8	15
Nail salon[b]	—	25
Pet shops	—	1.00 cfm/ft^2
Reducing salons	20	15
Supermarkets	8	15

(continued)

TABLE 403.3—continued
REQUIRED OUTDOOR VENTILATION AIR

OCCUPANCY CLASSIFICATION	ESTIMATED MAXIMUM OCCUPANCY LOAD, PERSONS PER 1,000 SQUARE FEET[a]	OUTDOOR AIR (Cubic feet per Minute (cfm) Per person) UNLESS NOTED[e]
Sports and amusement		
Ballrooms and discos	100	25
Bowling alleys (seating areas)	70	25
Game rooms	70	25
Ice arenas	—	0.50 cfm/ft²
Playing floors (gymnasiums)	30	20
Spectator areas	150	15
Swimming pools (pool and deck area)	—	0.50 cfm/ft²
Storage		
Repair garages, enclosed parking garages[d]	—	1.5 cfm/ft²
Warehouses	—	0.05 cfm/ft²
Theaters		
Auditoriums	150	15
Lobbies	150	20
Stages, studios	70	15
Ticket booths	60	20
Transportation		
Platforms	100	15
Vehicles	150	15
Waiting rooms	100	15
Workrooms		
Bank vaults	5	15
Darkrooms	—	0.50 cfm/ft²
Duplicating, printing	—	0.50 cfm/ft²
Meat processing[c]	10	15
Pharmacy	20	15
Photo studios	10	15

For SI: 1 cubic foot per minute = 0.0004719 m³/s, 1 ton = 908 kg,

1 cubic foot per minute per square foot = 0.00508 m³/(s • m²),

°C = [(°F) -32]/1.8, 1 square foot = 0.0929 m².

a. Based upon net floor area.

b. Mechanical exhaust required and the recirculation of air from such spaces as permitted by Section 403.2.1 is prohibited (see Section 403.2.1).

c. Spaces unheated or maintained below 50°F are not covered by these requirements unless the occupancy is continuous.

d. Ventilation systems in enclosed parking garages shall comply with Section 404. A mechanical ventilation system shall not be required in garages having a floor area not exceeding 850 square feet and used for the storage of not more than four vehicles or trucks of 1 ton maximum capacity.

e. Where the ventilation rate is expressed in cfm/ft², such rate is based upon cubic feet per minute per square foot of the floor area being ventilated.

f. The sum of the outdoor and transfer air from adjacent spaces shall be sufficient to provide an exhaust rate of not less than 1.5 cfm/ft².

g. Transfer air permitted in accordance with Section 403.2.2.

403.3.3 Variable air volume system control. Variable air volume air distribution systems, other than those designed to supply only 100-percent outdoor air, shall be provided with controls to regulate the flow of outdoor air. Such control system shall be designed to maintain the flow of outdoor air at a rate of not less than that required by Section 403 over the entire range of supply air operating rates.

403.3.4 Balancing. Ventilation systems shall be balanced by an approved method. Such balancing shall verify that the ventilation system is capable of supplying the airflow rates required by Section 403.

SECTION 404
ENCLOSED PARKING GARAGES

404.1 Enclosed parking garages. Mechanical ventilation systems for enclosed parking garages are not required to operate continuously where the system is arranged to operate automatically upon detection of a concentration of carbon monoxide of 25 parts per million (ppm) by approved automatic detection devices.

404.2 Minimum ventilation. Automatic operation of the system shall not reduce the ventilation rate below 0.05 cfm per square foot (0.00025 m³/s • m²) of the floor area and the system shall be capable of producing a ventilation rate of 1.5 cfm per square foot (0.0076m³/s • m²) of floor area.

404.3 Occupied spaces accessory to public garages. Connecting offices, waiting rooms, ticket booths and similar uses that are accessory to a public garage shall be maintained at a positive pressure and shall be provided with ventilation in accordance with Section 403.3.

SECTION 405
SYSTEMS CONTROL

405.1 General. Mechanical ventilation systems shall be provided with manual or automatic controls that will operate such systems whenever the spaces are occupied. Air-conditioning systems that supply required ventilation air shall be provided with controls designed to automatically maintain the required outdoor air supply rate during occupancy.

SECTION 406
VENTILATION OF UNINHABITED SPACES

406.1 General. Uninhabited spaces, such as crawl spaces and attics, shall be provided with natural ventilation openings as required by the *International Building Code* or shall be provided with a mechanical exhaust and supply air system. The mechanical exhaust rate shall be not less than 0.02 cfm per square foot (0.00001 m³/s · m²) of horizontal area and shall be automatically controlled to operate when the relative humidity in the space served exceeds 60 percent.

CHAPTER 5
EXHAUST SYSTEMS

SECTION 501
GENERAL

501.1 Scope. This chapter shall govern the design, construction and installation of mechanical exhaust systems, including dust, stock and refuse conveyor systems, exhaust systems serving commercial cooking appliances and energy recovery ventilation systems.

501.2 Outdoor discharge. The air removed by every mechanical exhaust system shall be discharged outdoors at a point where it will not cause a nuisance and from which it cannot again be readily drawn in by a ventilating system. Air shall not be exhausted into an attic or crawl space.

Exceptions:

1. Whole-house ventilation-type attic fans that discharge into the attic space of dwelling units having private attics.

2. Commercial cooking recirculating systems.

501.3 Pressure equalization. Mechanical exhaust systems shall be sized to remove the quantity of air required by this chapter to be exhausted. The system shall operate when air is required to be exhausted. Where mechanical exhaust is required in a room or space in other than occupancies in Group R-3, such space shall be maintained with a neutral or negative pressure. If a greater quantity of air is supplied by a mechanical ventilating supply system than is removed by a mechanical exhaust system for a room, adequate means shall be provided for the natural exit of the excess air supplied. If only a mechanical exhaust system is installed for a room or if a greater quantity of air is removed by a mechanical exhaust system than is supplied by a mechanical ventilating supply system for a room, adequate means shall be provided for the natural supply of the deficiency in the air supplied.

501.4 Ducts. Where exhaust duct construction is not specified in this chapter, such construction shall comply with Chapter 6.

SECTION 502
REQUIRED SYSTEMS

502.1 General. An exhaust system shall be provided, maintained and operated as specifically required by this section and for all occupied areas where machines, vats, tanks, furnaces, forges, salamanders and other appliances, equipment and processes in such areas produce or throw off dust or particles sufficiently light to float in the air, or which emit heat, odors, fumes, spray, gas or smoke, in such quantities so as to be irritating or injurious to health or safety.

502.1.1 Exhaust location. The inlet to an exhaust system shall be located in the area of heaviest concentration of contaminants.

[F] 502.1.2 Fuel-dispensing areas. The bottom of an air inlet or exhaust opening in fuel-dispensing areas shall be located not more than 18 inches (457 mm) above the floor.

502.1.3 Equipment, appliance and service rooms. Equipment, appliance and system service rooms that house sources of odors, fumes, noxious gases, smoke, steam, dust, spray or other contaminants shall be designed and constructed so as to prevent spreading of such contaminants to other occupied parts of the building.

[F] 502.1.4 Hazardous exhaust. The mechanical exhaust of high concentrations of dust or hazardous vapors shall conform to the requirements of Section 510.

[F] 502.2 Aircraft fueling and defueling. Compartments housing piping, pumps, air eliminators, water separators, hose reels and similar equipment used in aircraft fueling and defueling operations shall be adequately ventilated at floor level or within the floor itself.

[F] 502.3 Battery-charging areas. Ventilation shall be provided in an approved manner in battery-charging areas to prevent a dangerous accumulation of flammable gases.

[F] 502.4 Stationary lead-acid battery systems. Ventilation shall be provided for stationary lead-acid battery systems in accordance with this chapter and Section 502.4.1 or 502.4.2.

[F] 502.4.1 Hydrogen limit. The ventilation system shall be designed to limit the maximum concentration of hydrogen to 1.0 percent of the total volume of the room.

[F] 502.4.2 Ventilation rate. Continuous ventilation shall be provided at a rate of not less than 1 cubic foot per minute per square foot (cfm/ft^2) [0.00508 m^3/(s $\cdot$ m^2)] of floor area of the room.

[F] 502.5 Valve-regulated lead-acid batteries. Valve-regulated lead-acid battery systems as regulated by Section 609 of the *International Fire Code*, shall be provided with ventilation in accordance with Section 502.5.1 or 502.5.2 for rooms and in accordance with Section 502.5.3 or 502.5.4 for cabinets.

[F] 502.5.1 Hydrogen limit in rooms. The ventilation system shall be designed to limit the maximum concentration of hydrogen to 1.0 percent of the total volume of the room during the worst-case event of simultaneous boost charging of all batteries in the room.

[F] 502.5.2 Ventilation rate in rooms. Continuous ventilation shall be provided at a rate of not less than 1 cubic foot per minute per square foot (cfm/ft.^{2}) [0.00508 m^3/(s$\cdot$m^2)] of floor area of the room.

[F] 502.5.3 Hydrogen limit in cabinets. The ventilation system shall be designed to limit the maximum concentration of hydrogen to 1.0 percent of the total volume of the cabinet during the worst-case event of simultaneous boost charging of all batteries in the cabinet.

[F]502.5.4 Ventilation rate in cabinets. Continuous ventilation shall be provided at a rate of not less than 1 cubic foot per minute per square foot (cfm/ft.²) [0.00508 m³/(s•m²)] of the floor area covered by the cabinet. The room in which the cabinet is installed shall also be ventilated as required by Section 502.5.1 or 502.5.2.

[F] 502.6 Dry cleaning plants. Ventilation in dry cleaning plants shall be adequate to protect employees and the public in accordance with this section and DOL 29 CFR Part 1910.1000, where applicable.

[F] 502.6.1 Type II systems. Type II dry cleaning systems shall be provided with a mechanical ventilation system that is designed to exhaust 1 cubic foot of air per minute for each square foot of floor area (1 cfm/ft²) [0.00508 m³/(s • m²)] in dry cleaning rooms and in drying rooms. The ventilation system shall operate automatically when the dry cleaning equipment is in operation and shall have manual controls at an approved location.

[F] 502.6.2 Type IV and V systems. Type IV and V dry cleaning systems shall be provided with an automatically activated exhaust ventilation system to maintain a minimum of 100 feet per minute (0.5 m/s) air velocity through the loading door when the door is opened.

Exception: Dry cleaning units are not required to be provided with exhaust ventilation where an exhaust hood is installed immediately outside of and above the loading door which operates at an airflow rate as follows:

$$Q = 100 \times A_{LD} \qquad \text{(Equation 5-1)}$$

where:

Q = Flow rate exhausted through the hood, cubic feet per minute.

A_{LD} = Area of the loading door, square feet.

[F] 502.6.3 Spotting and pretreating. Scrubbing tubs, scouring, brushing or spotting operations shall be located such that solvent vapors are captured and exhausted by the ventilating system.

[F] 502.7 Application of flammable finishes. Mechanical exhaust as required by this section shall be provided for operations involving the application of flammable finishes.

[F] 502.7.1 During construction. Ventilation shall be provided for operations involving the application of materials containing flammable solvents in the course of construction, alteration or demolition of a structure.

[F] 502.7.2 Limited spraying spaces. Positive mechanical ventilation which provides a minimum of six complete air changes per hour shall be installed in limited spraying spaces. Such system shall meet the requirements of the *International Fire Code* for handling flammable vapors. Explosion venting is not required.

[F] 502.7.3 Spraying areas. Mechanical ventilation of spraying areas shall be provided in accordance with Sections 502.7.3.1 through 502.7.3.7.

[F] 502.7.3.1 Operation. Mechanical ventilation shall be kept in operation at all times while spraying operations are being conducted and for a sufficient time thereafter to al-low vapors from drying coated articles and finishing material residue to be exhausted. Spraying equipment shall be interlocked with the ventilation of the spraying area such that spraying operations cannot be conducted unless the ventilation system is in operation.

[F] 502.7.3.2 Recirculation. Air exhausted from spraying operations shall not be recirculated.

Exceptions:

1. Air exhausted from spraying operations shall be permitted to be recirculated as makeup air for un-manned spray operations provided that:

   1.1. Solid particulate has been removed.

   1.2. The vapor concentration is less than 25 percent of the lower flammablity limit (LFL).

   1.3. Approved equipment is used to monitor the vapor concentration.

   1.4. An alarm is sounded and spray operations are automatically shut down if the vapor concentration exceeds 25 percent of the LFL.

   1.5. The spray booths, spray spaces or spray rooms involved in any recirculation process shall be provided with mechanical ventilation that shall automatically exhaust 100 percent of the required air volume in the event of shutdown by approved equipment used to monitor vapor concentrations.

2. Air exhausted from spraying operations shall be permitted to be recirculated as makeup air to manned spraying operations if all of the conditions provided in Exception 1 are included in the installation and documents have been prepared to show that the installation does not present life safety hazards to personnel inside the spray booth, spray space or spray room.

[F] 502.7.3.3 Air velocity. Ventilation systems shall be designed, installed and maintained such that the average air velocity over the open face of the booth, or booth cross section in the direction of airflow during spraying operations, is not less than 100 feet per minute (0.51 m/s).

[F] 502.7.3.4 Ventilation obstruction. Articles being sprayed shall be positioned in a manner that does not obstruct collection of overspray.

[F] 502.7.3.5 Independent ducts. Each spray booth and spray room shall have an independent exhaust duct system discharging to the outdoors.

Exceptions:

1. Multiple spray booths having a combined frontal area of 18 square feet (1.67 m²) or less are allowed to have a common exhaust where identical spray-finishing material is used in each booth. If more than one fan serves one booth, such fans shall be interconnected so that all fans operate simultaneously.

2. Where treatment of exhaust is necessary for air pollution control or energy conservation, ducts shall be allowed to be manifolded if all of the following conditions are met:

   2.1. The sprayed materials used are compatible and will not react or cause ignition of the residue in the ducts.

   2.2. Nitrocellulose-based finishing material shall not be used.

   2.3. A filtering system shall be provided to reduce the amount of overspray carried into the duct manifold.

   2.4. Automatic sprinkler protection shall be provided at the junction of each booth exhaust with the manifold, in addition to the protection required by this chapter.

[F] 502.7.3.6 Termination point. The termination point for exhaust ducts discharging to the atmosphere shall be located with the following minimum distances.

1. For ducts conveying explosive or flammable vapors, fumes or dusts: 30 feet (9144 mm) from the property line; 10 feet (3048 mm) from openings into the building; 6 feet (1829 mm) from exterior walls and roofs; 30 feet (9144 mm) from combustible walls and openings into the building which are in the direction of the exhaust discharge; 10 feet (3048 mm) above adjoining grade.

2. For other product-conveying outlets: 10 feet (3048 mm) from the property line; 3 feet (914 mm) from exterior walls and roofs; 10 feet (3048 mm) from openings into the building; 10 feet (3048 mm) above adjoining grade.

3. For environmental air duct exhaust: 3 feet (914 mm) from the property line; 3 feet (914 mm) from openings into the building.

[F] 502.7.3.7 Fan motors and belts. Electric motors driving exhaust fans shall not be placed inside booths or ducts. Fan rotating elements shall be nonferrous or nonsparking or the casing shall consist of, or be lined with, such material. Belts shall not enter the duct or booth unless the belt and pulley within the duct are tightly enclosed.

[F] 502.7.4 Dipping operations. Vapor areas of dip tank operations shall be provided with mechanical ventilation adequate to prevent the dangerous accumulation of vapors. Required ventilation systems shall be so arranged that the failure of any ventilating fan will automatically stop the dipping conveyor system.

[F] 502.7.5 Electrostatic apparatus. The spraying area in spray-finishing operations involving electrostatic apparatus and devices shall be ventilated in accordance with Section 502.7.3.

[F] 502.7.6 Powder coating. Exhaust ventilation for powder-coating operations shall be sufficient to maintain the atmosphere below one-half of the minimum explosive concentration for the material being applied. Nondeposited, air-suspended powders shall be removed through exhaust ducts to the powder recovery cyclone or receptacle.

[F] 502.7.7 Floor resurfacing operations. To prevent the accumulation of flammable vapors during floor resurfacing operations, mechanical ventilation at a minimum rate of 1 cfm/ft² [0.00508 m³/(s · m²)] of area being finished shall be provided. Such exhaust shall be by approved temporary or portable means. Vapors shall be exhausted to the exterior of the building.

[F] 502.8 Hazardous materials—general requirements. Exhaust ventilation systems for structures containing hazardous materials shall be provided as required in Sections 502.8.1 through 502.8.5.

[F] 502.8.1 Storage in excess of the maximum allowable quantities. Indoor storage areas and storage buildings for hazardous materials in amounts exceeding the maximum allowable quantity per control area shall be provided with mechanical exhaust ventilation or natural ventilation where natural ventilation can be shown to be acceptable for the materials as stored.

> **Exception:** Storage areas for flammable solids complying with the *International Fire Code*.

[F] 502.8.1.1 System requirements. Exhaust ventilation systems shall comply with all of the following:

1. The installation shall be in accordance with this code.

2. Mechanical ventilation shall be provided at a rate of not less than 1 cfm/ft² [0.00508 m³/(s · m²)] of floor area over the storage area.

3. The systems shall operate continuously unless alternate designs are approved.

4. A manual shutoff control shall be provided outside of the room in a position adjacent to the access door to the room or in another approved location. The switch shall be of the break-glass type and shall be labeled: VENTILATION SYSTEM EMERGENCY SHUTOFF.

5. The exhaust ventilation system shall be designed to consider the density of the potential fumes or vapors released. For fumes or vapors that are heavier than air, exhaust shall be taken from a point within 12 inches (304 mm) of the floor.

6. The location of both the exhaust and inlet air openings shall be designed to provide air movement across all portions of the floor or room to prevent the accumulation of vapors.

7. The exhaust ventilation shall not be recirculated within the room or building if the materials stored are capable of emitting hazardous vapors.

[F] 502.8.2 Gas rooms, exhausted enclosures and gas cabinets. The ventilation system for gas rooms, exhausted enclosures and gas cabinets for any quantity of hazardous material shall be designed to operate at a negative pressure in relation to the surrounding area. Highly toxic and toxic gases shall also comply with Sections 502.9.7.1, 502.9.7.2 and 502.9.8.4.

[F] 502.8.3 Indoor dispensing and use. Indoor dispensing and use areas for hazardous materials in amounts exceeding the maximum allowable quantity per control area shall be provided with exhaust ventilation in accordance with Section 502.7.1.

> **Exception:** Ventilation is not required for dispensing and use of flammable solids other than finely divided particles.

[F] 502.8.4 Indoor dispensing and use—point sources. Where gases, liquids or solids in amounts exceeding the maximum allowable quantity per control area and having a hazard ranking of 3 or 4 in accordance with NFPA 704 are dispensed or used, mechanical exhaust ventilation shall be provided to capture fumes, mists or vapors at the point of generation.

> **Exception:** Where it can be demonstrated that the gases, liquids or solids do not create harmful fumes, mists or vapors.

[F] 502.8.5 Closed systems. Where closed systems for the use of hazardous materials in amounts exceeding the maximum allowable quantity per control area are designed to be opened as part of normal operations, ventilation shall be provided in accordance with Section 502.8.4.

[F] 502.9 Hazardous materials—requirements for specific materials. Exhaust ventilation systems for specific hazardous materials shall be provided as required in Section 502.8 and Sections 502.9.1 through 502.9.11.

[F] 502.9.1 Compressed gases—medical gas systems. Rooms for the storage of compressed medical gases in amounts exceeding the maximum allowable exempt quantity per control area, and which do not have an exterior wall, shall be exhausted through a duct to the exterior of the building. Both separate airstreams shall be enclosed in a 1-hour-rated shaft enclosure from the room to the exterior. Approved mechanical ventilation shall be provided at a minimum rate of 1 cfm/ft^2 [0.00508 m^3/(s · m^2)] of the area of the room.

Gas cabinets for the storage of compressed medical gases in amounts exceeding the maximum allowable quantity per control area shall be connected to an exhaust system. The average velocity of ventilation at the face of access ports or windows shall be not less than 200 feet per minute (1.02 m/s) with a minimum velocity of 150 feet per minute (0.76 m/s) at any point at the access port or window.

[F] 502.9.2 Corrosives. Where corrosive materials in amounts exceeding the maximum allowable quantity per control area are dispensed or used, mechanical exhaust ventilation in accordance with Section 502.8.4 shall be provided.

[F] 502.9.3 Cryogenics. Storage areas for stationary or portable containers of cryogenic fluids in any quantity shall be ventilated in accordance with Section 502.8. Indoor areas where cryogenic fluids in any quantity are dispensed shall be ventilated in accordance with the requirements of Section 502.8.4 in a manner that captures any vapor at the point of generation.

> **Exception:** Ventilation for indoor dispensing areas is not required where it can be demonstrated that the cryogenic fluids do not create harmful vapors.

[F] 502.9.4 Explosives. Squirrel cage blowers shall not be used for exhausting hazardous fumes, vapors or gases in operating buildings and rooms for the manufacture, assembly or testing of explosives. Only nonferrous fan blades shall be used for fans located within the ductwork and through which hazardous materials are exhausted. Motors shall be located outside the duct.

[F] 502.9.5 Flammable and combustible liquids. Exhaust ventilation systems shall be provided as required by Sections 502.9.5.1 through 502.9.5.5 for the storage, use, dispensing, mixing and handling of flammable and combustible liquids. Unless otherwise specified, this section shall apply to any quantity of flammable and combustible liquids.

> **Exception:** This section shall not apply to flammable and combustible liquids that are exempt from the *International Fire Code.*

[F] 502.9.5.1 Vaults. Vaults that contain tanks of Class I liquids shall be provided with continuous ventilation at a rate of not less than 1 cfm/ft^2 of floor area [0.00508 m^3/(s · m^2)], but not less than 150 cfm (4 m^3/min). Failure of the exhaust airflow shall automatically shut down the dispensing system. The exhaust system shall be designed to provide air movement across all parts of the vault floor. Supply and exhaust ducts shall extend to a point not greater than 12 inches (305 mm) and not less than 3 inches (76 mm) above the floor. The exhaust system shall be installed in accordance with the provisions of NFPA 91. Means shall be provided to automatically detect any flammable vapors and to automatically shut down the dispensing system upon detection of such flammable vapors in the exhaust duct at a concentration of 25 percent of the LFL.

[F] 502.9.5.2 Storage rooms and warehouses. Liquid storage rooms and liquid storage warehouses for quantities of liquids exceeding those specified in the *International Fire Code* shall be ventilated in accordance with Section 502.8.1.

[F] 502.9.5.3 Cleaning machines. Areas containing machines used for parts cleaning in accordance with the *International Fire Code* shall be adequately ventilated to prevent accumulation of vapors.

[F] 502.9.5.4 Use, dispensing and mixing. Continuous mechanical ventilation shall be provided for the use, dispensing and mixing of flammable and combustible liquids in open or closed systems in amounts exceeding the maximum allowable quantity per control area and for bulk transfer and process transfer operations. The ventilation rate shall be not less than 1 cfm/ft^2 [0.00508 m^3/(s · m^2)] of floor area over the design area. Provisions shall be made for the introduction of makeup air in a manner that will include all floor areas or pits where vapors can collect. Local or spot ventilation shall be provided where needed to prevent the accumulation of hazardous vapors.

> **Exception:** Where natural ventilation can be shown to be effective for the materials used, dispensed, or mixed.

[F] 502.9.5.5 Bulk plants or terminals. Ventilation shall be provided for portions of properties where flammable and combustible liquids are received by tank vessels, pipelines, tank cars or tank vehicles and which are stored or blended in bulk for the purpose of distributing such liquids by tank vessels, pipelines, tank cars, tank vehicles or containers as required by Sections 502.9.5.5.1 through 502.9.5.5.3.

[F] 502.9.5.5.1 General. Ventilation shall be provided for rooms, buildings and enclosures in which Class I liquids are pumped, used or transferred. Design of ventilation systems shall consider the relatively high specific gravity of the vapors. Where natural ventilation is used, adequate openings in outside walls at floor level, unobstructed except by louvers or coarse screens, shall be provided. Where natural ventilation is inadequate, mechanical ventilation shall be provided.

[F] 502.9.5.5.2 Basements and pits. Class I liquids shall not be stored or used within a building having a basement or pit into which flammable vapors can travel, unless such area is provided with ventilation designed to prevent the accumulation of flammable vapors therein.

[F] 502.9.5.5.3 Dispensing of Class I liquids. Containers of Class I liquids shall not be drawn from or filled within buildings unless a provision is made to prevent the accumulation of flammable vapors in hazardous concentrations. Where mechanical ventilation is required, it shall be kept in operation while flammable vapors could be present.

[F] 502.9.6 Highly toxic and toxic liquids. Ventilation exhaust shall be provided for highly toxic and toxic liquids as required by Sections 502.9.6.1 and 502.9.6.2.

[F] 502.9.6.1 Treatment system. This provision shall apply to indoor and outdoor storage and use of highly toxic and toxic liquids in amounts exceeding the maximum allowable quantities per control area. Exhaust scrubbers or other systems for processing vapors of highly toxic liquids shall be provided where a spill or accidental release of such liquids can be expected to release highly toxic vapors at normal temperature and pressure.

[F] 502.9.6.2 Open and closed systems. Mechanical exhaust ventilation shall be provided for highly toxic and toxic liquids used in open systems in accordance with Section 502.8.4. Mechanical exhaust ventilation shall be provided for highly toxic and toxic liquids used in closed systems in accordance with Section 502.8.5.

Exception: Liquids or solids that do not generate highly toxic or toxic fumes, mists or vapors.

[F] 502.9.7 Highly toxic and toxic compressed gases— any quantity. Ventilation exhaust shall be provided for highly toxic and toxic compressed gases in any quantity as required by Sections 502.9.7.1 and 502.9.7.2.

[F] 502.9.7.1 Gas cabinets. Gas cabinets containing highly toxic or toxic compressed gases in any quantity shall comply with Section 502.8.2 and the following requirements:

1. The average ventilation velocity at the face of gas cabinet access ports or windows shall be not less than 200 feet per minute (1.02 m/s) with a minimum velocity of 150 feet per minute (0.76 m/s) at any point at the access port or window.

2. Gas cabinets shall be connected to an exhaust system.

3. Gas cabinets shall not be used as the sole means of exhaust for any room or area.

[F] 502.9.7.2 Exhausted enclosures. Exhausted enclosures containing highly toxic or toxic compressed gases in any quantity shall comply with Section 502.8.2 and the following requirements:

1. The average ventilation velocity at the face of the enclosure shall be not less than 200 feet per minute (1.02 m/s) with a minimum velocity of 150 feet per minute (0.76 m/s).

2. Exhausted enclosures shall be connected to an exhaust system.

3. Exhausted enclosures shall not be used as the sole means of exhaust for any room or area.

[F] 502.9.8 Highly toxic and toxic compressed gases— quantities exceeding the maximum allowable per control area. Ventilation exhaust shall be provided for highly toxic and toxic compressed gases in amounts exceeding the maximum allowable quantities per control area as required by Sections 502.9.8.1 through 502.9.8.6.

[F] 502.9.8.1 Ventilated areas. The room or area in which indoor gas cabinets or exhausted enclosures are located shall be provided with exhaust ventilation. Gas cabinets or exhausted enclosures shall not be used as the sole means of exhaust for any room or area.

[F] 502.9.8.2 Local exhaust for portable tanks. A means of local exhaust shall be provided to capture leakage from indoor and outdoor portable tanks. The local exhaust shall consist of portable ducts or collection systems designed to be applied to the site of a leak in a valve or fitting on the tank. The local exhaust system shall be located in a gas room. Exhaust shall be directed to a treatment system where required by the *International Fire Code*.

[F] 502.9.8.3 Piping and controls—stationary tanks. Filling or dispensing connections on indoor stationary tanks shall be provided with a means of local exhaust. Such exhaust shall be designed to capture fumes and vapors. The exhaust shall be directed to a treatment system where required by the *International Fire Code*.

[F] 502.9.8.4 Gas rooms. The ventilation system for gas rooms shall be designed to operate at a negative pressure in relation to the surrounding area. The exhaust ventilation from gas rooms shall be directed to an exhaust system.

[F] 502.9.8.5 Treatment system. The exhaust ventilation from gas cabinets, exhausted enclosures and gas rooms, and local exhaust systems required in Sections 502.9.8.2 and 502.9.8.3 shall be directed to a treatment system where required by the *International Fire Code*.

[F] 502.9.8.6 Process equipment. Effluent from indoor and outdoor process equipment containing highly toxic or toxic compressed gases which could be discharged to the atmosphere shall be processed through an exhaust scrubber or other processing system. Such systems shall be in accordance with the *International Fire Code*.

[F] 502.9.9 Ozone gas generators. Ozone cabinets and ozone gas-generator rooms for systems having a maximum ozone-generating capacity of one-half pound (0.23 kg) or more over a 24-hour period shall be mechanically ventilated at a rate of not less than six air changes per hour. For cabinets, the average velocity of ventilation at makeup air openings with cabinet doors closed shall be not less than 200 feet per minute (1.02 m/s).

[F] 502.9.10 LP-gas distribution facilities. LP-gas distribution facilities shall be ventilated in accordance with NFPA 58.

[F] 502.9.10.1 Portable container use. Above-grade underfloor spaces or basements in which portable LP-gas containers are used or are stored awaiting use or re-sale shall be provided with an approved means of ventilation.

Exception: Department of Transportation (DOT) specification cylinders with a maximum water capacity of 2.5 pounds (1 kg) for use in completely self-contained hand torches and similar applications. The quantity of LP-gas shall not exceed 20 pounds (9 kg).

[F] 502.9.11 Silane gas. Exhausted enclosures and gas cabinets for the indoor storage of silane gas in amounts exceeding the maximum allowable quantities per control area shall comply with this section.

1. Exhausted enclosures and gas cabinets shall be in accordance with Section 502.8.2.

2. The velocity of ventilation across unwelded fittings and connections on the piping system shall not be less than 200 feet per minute (1.02 m/s).

3. The average velocity at the face of the access ports or windows in the gas cabinet shall not be less than 200 feet per minute (1.02 m/s) with a minimum velocity of 150 feet per minute (0.76 m/s) at any point at the access port or window.

[F] 502.10 Hazardous production materials (HPM). Exhaust ventilation systems and materials for ducts utilized for the exhaust of HPM shall comply with this section, other applicable provisions of this code, the *International Building Code* and the *International Fire Code*.

[F] 502.10.1 Where required. Exhaust ventilation systems shall be provided in the following locations in accordance with the requirements of this section and the *International Building Code*:

1. Fabrication areas: Exhaust ventilation for fabrication areas shall comply with the *International Building Code*. Additional manual control switches shall be provided where required by the code official.

2. Workstations: A ventilation system shall be provided to capture and exhaust fumes and vapors at workstations.

3. Liquid storage rooms: Exhaust ventilation for liquid storage rooms shall comply with Section 502.8.1.1 and the *International Building Code*.

4. HPM rooms: Exhaust ventilation for HPM rooms shall comply with Section 502.8.1.1 and the *International Building Code*.

5. Gas cabinets: Exhaust ventilation for gas cabinets shall comply with Section 502.8.2. The gas cabinet ventilation system is allowed to connect to a workstation ventilation system. Exhaust ventilation for gas cabinets containing highly toxic or toxic gases shall also comply with Sections 502.9.7 and 502.9.8.

6. Exhausted enclosures: Exhaust ventilation for exhausted enclosures shall comply with Section 502.7.2. Exhaust ventilation for exhausted enclosures containing highly toxic or toxic gases shall also comply with Sections 502.9.7 and 502.9.8.

7. Gas rooms: Exhaust ventilation for gas rooms shall comply with Section 502.8.2. Exhaust ventilation for gas cabinets containing highly toxic or toxic gases shall also comply with Sections 502.9.7 and 502.9.8.

[F] 502.10.2 Penetrations. Exhaust ducts penetrating fire barrier assemblies shall be contained in a shaft of equivalent fire-resistive construction. Exhaust ducts shall not penetrate building separation walls. Fire dampers shall not be installed in exhaust ducts.

[F] 502.10.3 Treatment systems. Treatment systems for highly toxic and toxic gases shall comply with the *International Fire Code*.

502.11 Motion picture projectors. Motion picture projectors shall be exhausted in accordance with Section 502.11.1 or 502.11.2.

502.11.1 Projectors with an exhaust discharge. Projectors equipped with an exhaust discharge shall be directly connected to a mechanical exhaust system. The exhaust system shall operate at an exhaust rate as indicated by the manufacturer's installation instructions.

502.11.2 Projectors without exhaust connection. Projectors without an exhaust connection shall have contaminants exhausted through a mechanical exhaust system. The exhaust rate for electric arc projectors shall be a minimum of 200 cubic feet per minute (cfm) (0.09 m³/s) per lamp. The exhaust rate for xenon projectors shall be a minimum of 300 cfm (0.14 m³/s) per lamp. Xenon projector exhaust shall be at a rate such that the exterior temperature of the lamp housing does not exceed 130°F (54°C). The lamp and projection room exhaust systems, whether combined or independent, shall not be interconnected with any other exhaust or return system within the building.

[F] 502.12 Organic coating processes. Enclosed structures involving organic coating processes in which Class I liquids are processed or handled shall be ventilated at a rate of not less than 1 cfm/ft² [0.00508 m³/(s · m²)] of solid floor area. Ventilation shall be accomplished by exhaust fans that intake at floor levels and discharge to a safe location outside the structure. Noncontaminated intake air shall be introduced in such a manner that all portions of solid floor areas are provided with continuous uniformly distributed air movement.

502.13 Public garages. Mechanical exhaust systems for public garages, as required in Chapter 4, shall operate continuously or in accordance with Section 404.

502.14 Motor vehicle operation. In areas where motor vehicles operate, mechanical ventilation shall be provided in accordance with Section 403. Additionally, areas in which stationary motor vehicles are operated shall be provided with a source capture system that connects directly to the motor vehicle exhaust systems.

Exceptions:

1. This section shall not apply where the motor vehicles being operated or repaired are electrically powered.

2. This section shall not apply to one- and two-family dwellings.

3. This section shall not apply to motor vehicle service areas where engines are operated inside the building only for the duration necessary to move the motor vehicles in and out of the building.

[F] 502.15 Repair garages. Where Class I liquids or LP-gas are stored or used within a building having a basement or pit wherein flammable vapors could accumulate, the basement or pit shall be provided with ventilation designed to prevent the accumulation of flammable vapors therein.

[F] 502.16 Repair garages for natural gas- and hydrogen-fueled vehicles. Repair garages used for the repair of natural gas- or hydrogen-fueled vehicles shall be provided with an approved mechanical ventilation system. The mechanical ventilation system shall be in accordance with Sections 502.16.1 and 502.16.2.

Exception: Where approved by the code official, natural ventilation shall be permitted in lieu of mechanical ventilation.

[F] 502.16.1 Design. Indoor locations shall be ventilated utilizing air supply inlets and exhaust outlets arranged to provide uniform air movement to the extent practical. Inlets shall be uniformly arranged on exterior walls near floor level. Outlets shall be located at the high point of the room in exterior walls or the roof.

1. Ventilation shall be by a continuous mechanical ventilation system or by a mechanical ventilation system activated by a continuously monitoring natural gas detection system activating at a gas concentration of 25 percent of the LFL. In all cases, the system shall shut down the fueling system in the event of failure of the ventilation system.

2. The ventilation rate shall be at least 1 cubic foot per minute per 12 cubic feet [$0.00138 \text{ m}^3/(\text{s} \cdot \text{m}^3)$] of room volume.

[F] 502.16.2 Operation. The mechanical ventilation system shall operate continuously.

Exceptions:

1. Mechanical ventilation systems that are interlocked with a gas detection system designed in accordance with the *International Fire Code*.

2. Mechanical ventilation systems in garages that are used only for the repair of vehicles fueled by liquid fuels or odorized gases, such as CNG, where the ventilation system is electrically interlocked with the lighting circuit.

502.17 Tire rebuilding or recapping. Each room where rubber cement is used or mixed, or where flammable or combustible solvents are applied, shall be ventilated in accordance with the applicable provisions of NFPA 91.

502.17.1 Buffing machines. Each buffing machine shall be connected to a dust-collecting system that prevents the accumulation of the dust produced by the buffing process.

502.18 Specific rooms. Specific rooms, including bathrooms, locker rooms, smoking lounges and toilet rooms, shall be exhausted in accordance with the ventilation requirements of Chapter 4.

SECTION 503
MOTORS AND FANS

503.1 General. Motors and fans shall be sized to provide the required air movement. Motors in areas that contain flammable vapors or dusts shall be of a type approved for such environments. A manually operated remote control installed at an approved location shall be provided to shut off fans or blowers in flammable vapor or dust systems. Electrical equipment and appliances used in operations that generate explosive or flammable vapors, fumes or dusts shall be interlocked with the ventilation system so that the equipment and appliances cannot be operated unless the ventilation fans are in operation. Motors for fans used to convey flammable vapors or dusts shall be located outside the duct or shall be protected with approved shields and dustproofing. Motors and fans shall be provided with a means of access for servicing and maintenance.

503.2 Fans. Parts of fans in contact with explosive or flammable vapors, fumes or dusts shall be of nonferrous or nonsparking materials, or their casing shall be lined or constructed of such material. When the size and hardness of materials passing through a fan are capable of producing a spark, both the fan and the casing shall be of nonsparking materials. When fans are required to be spark resistant, their bearings shall not be within the airstream, and all parts of the fan shall be grounded. Fans in systems-handling materials that are capable of clogging the blades, and fans in buffing or woodworking exhaust systems, shall be of the radial-blade or tube-axial type.

503.3 Equipment and appliances identification plate. Equipment and appliances used to exhaust explosive or flammable vapors, fumes or dusts shall bear an identification plate stating the ventilation rate for which the system was designed.

503.4 Corrosion-resistant fans. Fans located in systems conveying corrosives shall be of materials that are resistant to the corrosive or shall be coated with corrosion-resistant materials.

SECTION 504
CLOTHES DRYER EXHAUST

504.1 Installation. Clothes dryers shall be exhausted in accordance with the manufacturer's instructions. Dryer exhaust systems shall be independent of all other systems and shall convey the moisture and any products of combustion to the outside of the building.

Exception: This section shall not apply to listed and labeled condensing (ductless) clothes dryers.

504.2 Exhaust penetrations. Ducts that exhaust clothes dryers shall not penetrate or be located within any fireblocking, draftstopping or any wall, floor/ceiling or other assembly required by the *International Building Code* to be fire-resistance rated, unless such duct is constructed of galvanized steel or aluminum of the thickness specified in Section 603.4 and the fire-resistance rating is maintained in accordance with the *International Building Code*. Fire dampers, combination fire/smoke dampers and any similar devices that will obstruct the exhaust flow, shall be prohibited in clothes dryer exhaust ducts.

504.3 Cleanout. Each vertical riser shall be provided with a means for cleanout.

504.4 Exhaust installation. Dryer exhaust ducts for clothes dryers shall terminate on the outside of the building and shall be equipped with a backdraft damper. Screens shall not be installed at the duct termination. Ducts shall not be connected or installed with sheet metal screws or other fasteners that will obstruct the exhaust flow. Clothes dryer exhaust ducts shall not be connected to a vent connector, vent or chimney. Clothes dryer exhaust ducts shall not extend into or through ducts or plenums.

504.5 Makeup air. Installations exhausting more than 200 cfm (0.09 m³/s) shall be provided with makeup air. Where a closet is designed for the installation of a clothes dryer, an opening having an area of not less than 100 square inches (0.0645 m²) shall be provided in the closet enclosure.

504.6 Domestic clothes dryer ducts. Exhaust ducts for domestic clothes dryers shall be constructed of metal and shall have a smooth interior finish. The exhaust duct shall be a minimum nominal size of 4 inches (102 mm) in diameter. The entire exhaust system shall be supported and secured in place. The male end of the duct at overlapped duct joints shall extend in the direction of airflow. Clothes dryer transition ducts used to connect the appliance to the exhaust duct system shall be limited to single lengths not to exceed 8 feet (2438 mm) and shall be listed and labeled for the application. Transition ducts shall not be concealed within construction.

504.6.1 Maximum length. The maximum length of a clothes dryer exhaust duct shall not exceed 25 ft (7620 mm) from the dryer location to the outlet terminal. The maximum length of duct shall be reduced 2½ feet for each 45-degree (0.79 rad) bend and 5 feet (1524mm) for each 90-degree (1.6 rad) bend. The maximum length of the exhaust duct does not include the transition duct.

Exception: Where the make and model of the clothes dryer to be installed is known and the manufacturer's installation instructions for such dryer are provided to the code official, the maximum length of the exhaust duct, including any transition duct, shall be permitted to be in accordance with the dryer manufacturer's installation instructions.

504.6.2 Rough-in required. Where a compartment or space for a domestic clothes dryer is provided, an exhaust duct system shall be installed in accordance with Sections 504.6 and 504.6.1.

504.7 Commercial clothes dryers. The installation of dryer exhaust ducts serving Type 2 clothes dryers shall comply with the appliance manufacturer's installation instructions. Exhaust

fan motors installed in exhaust systems shall be located outside of the airstream. In multiple installations, the fan shall operate continuously or be interlocked to operate when any individual unit is operating. Ducts shall have a minimum clearance of 6 inches (152 mm) to combustible materials. Clothes dryer transition ducts used to connect the appliance to the exhaust duct system shall be limited to single lengths not to exceed 8 feet (2438 mm) in length and shall be listed and labeled for the application. Transition ducts shall not be concealed within construction.

SECTION 505
DOMESTIC KITCHEN EXHAUST EQUIPMENT

505.1 Domestic systems. Where domestic range hoods and domestic appliances equipped with downdraft exhaust are located within dwelling units, such hoods and appliances shall discharge to the outdoors through ducts constructed of galvanized steel, stainless steel, aluminum or copper. Such ducts shall have smooth inner walls and shall be air tight and equipped with a backdraft damper.

Exceptions:

1. Where installed in accordance with the manufacturer's installation instructions and where mechanical or natural ventilation is otherwise provided in accordance with Chapter 4, listed and labeled ductless range hoods shall not be required to discharge to the outdoors.

2. Ducts for domestic kitchen cooking appliances equipped with downdraft exhaust systems shall be permitted to be constructed of Schedule 40 PVC pipe provided that the installation complies with all of the following:

   2.1. The duct shall be installed under a concrete slab poured on grade.

   2.2. The underfloor trench in which the duct is installed shall be completely backfilled with sand or gravel.

   2.3. The PVC duct shall extend not greater than 1 inch (25 mm) above the indoor concrete floor surface.

   2.4. The PVC duct shall extend not greater than 1 inch (25 mm) above grade outside of the building.

   2.5. The PVC ducts shall be solvent cemented.

SECTION 506
COMMERCIAL KITCHEN HOOD VENTILATION SYSTEM DUCTS AND EXHAUST EQUIPMENT

506.1 General. Commercial kitchen hood ventilation ducts and exhaust equipment shall comply with the requirements of this section. Commercial kitchen grease ducts shall be designed for the type of cooking appliance and hood served.

506.2 Corrosion protection. Ducts exposed to the outside atmosphere or subject to a corrosive environment shall be protected against corrosion in an approved manner.

506.3 Ducts serving Type I hoods. Type I exhaust ducts shall be independent of all other exhaust systems except as provided in

Section 506.3.5. Commercial kitchen duct systems serving Type I hoods shall be designed, constructed and installed in accordance with Sections 506.3.1 through 506.3.12.3.

506.3.1 Duct materials. Ducts serving Type I hoods shall be constructed of materials in accordance with Sections 506.3.1.1 and 506.3.1.2.

506.3.1.1 Grease duct materials. Grease ducts serving Type I hoods shall be constructed of steel not less than 0.055 inch (1.4 mm) (No. 16 Gage) in thickness or stainless steel not less than 0.044 inch (1.1 mm) (No. 18 Gage) in thickness.

Exception: Listed and labeled factory-built commercial kitchen grease ducts shall be installed in accordance with Section 304.1.

506.3.1.2 Makeup air ducts. Make up air ducts connecting to or within 18 inches (457 mm) of a Type I hood shall be constructed and installed in accordance with Sections 603.1, 603.3, 603.4, 603.9, 603.10 and 603.12. Duct insulation installed within 18 inches (457 mm) of a Type I hood shall be noncombustible or shall be listed for the application.

506.3.2 Joints, seams and penetrations of grease ducts. Joints, seams and penetrations of grease ducts shall be made with a continuous liquid-tight weld or braze made on the external surface of the duct system.

Exceptions:

1. Penetrations shall not be required to be welded or brazed where sealed by devices that are listed for the application.
2. Internal welding or brazing shall not be prohibited provided that the joint is formed or ground smooth and is provided with ready access for inspection.
3. Listed and labeled factory-built commercial kitchen grease ducts installed in accordance with Section 304.1.

506.3.2.1 Duct joint types. Duct joints shall be butt joints or overlapping duct joints of either the telescoping or bell type. Overlapping joints shall be installed to prevent ledges and obstructions from collecting grease or interfering with gravity drainage to the intended collection point. The difference between the inside cross-sectional dimensions of overlapping sections of duct shall not exceed 0.25 inch (6 mm). The length of overlap for overlapping duct joints shall not exceed 2 inches (51 mm).

506.3.2.2 Duct-to-hood joints. Duct-to-hood joints shall be made with continuous internal or external liquid-tight welded or brazed joints. Such joints shall be smooth, accessible for inspection, and without grease traps.

Exceptions: This section shall not apply to:

1. A vertical duct-to-hood collar connection made in the top plane of the hood in accordance with all of the following:

1.1. The hood duct opening shall have a 1-inch-deep (25 mm), full perimeter, welded flange turned down into the

hood interior at an angle of 90 degrees from the plane of the opening.

1.2. The duct shall have a 1-inch-deep (25 mm) flange made by a 1-inch by 1-inch (25 mm by 25 mm) angle iron welded to the full perimeter of the duct not less than 1 inch (25 mm) above the bottom end of the duct.

1.3. A gasket rated for use at not less than 1,500°F (815°C) is installed between the duct flange and the top of the hood.

1.4. The duct-to-hood joint shall be secured by stud bolts not less than 0.25 inch (6.4 mm) in diameter welded to the hood with a spacing not greater than 4 inches (102 mm) on center for the full perimeter of the opening. All bolts and nuts are to be secured with lockwashers.

2. Listed and labeled duct-to-hood collar connections installed in accordance with Section 304.1.

506.3.2.3 Duct-to-exhaust fan connections. Duct-to-exhaust fan connections shall be flanged and gasketed at the base of the fan for vertical discharge fans; shall be flanged, gasketed and bolted to the inlet of the fan for side-inlet utility fans; and shall be flanged, gasketed and bolted to the inlet and outlet of the fan for in-line fans.

506.3.2.4 Vibration isolation. A vibration isolation connector for connecting a duct to a fan shall consist of noncombustible packing in a metal sleeve joint of approved design or shall be a coated-fabric flexible duct connector listed and labeled for the application. Vibration isolation connectors shall be installed only at the connection of a duct to a fan inlet or outlet.

506.3.3 Grease duct supports. Grease duct bracing and supports shall be of noncombustible material securely attached to the structure and designed to carry gravity and seismic loads within the stress limitations of the *International Building Code*. Bolts, screws, rivets and other mechanical fasteners shall not penetrate duct walls.

506.3.4 Air velocity. Grease duct systems serving a Type I hood shall be designed and installed to provide an air velocity within the duct system of not less than 1,500 feet per minute (7.6 m/s).

Exception: The velocity limitations shall not apply within duct transitions utilized to connect ducts to differently sized or shaped openings in hoods and fans, provided that such transitions do not exceed 3 feet (914 mm) in length and are designed to prevent the trapping of grease.

506.3.5 Separation of grease duct system. A separate grease duct system shall be provided for each Type I hood. A separate grease duct system is not required where all of the following conditions are met:

1. All interconnected hoods are located within the same story.
2. All interconnected hoods are located within the same room or in adjoining rooms.

3. Interconnecting ducts do not penetrate assemblies required to be fire-resistance rated.

4. The grease duct system does not serve solid fuel-fired appliances.

506.3.6 Grease duct clearances. Grease duct systems and exhaust equipment serving a Type I hood shall have a clearance to combustible construction of not less than 18 inches (457 mm), and shall have a clearance to noncombustible construction and gypsum wallboard attached to noncombustible structures of not less than 3 inches (76 mm).

> **Exception:** Listed and labeled factory-built commercial kitchen grease ducts and exhaust equipment installed in accordance with Section 304.1.

506.3.7 Prevention of grease accumulation in grease ducts. Duct systems serving a Type I hood shall be constructed and installed so that grease cannot collect in any portion thereof, and the system shall slope not less than one-fourth unit vertical in 12 units horizontal (2-percent slope) toward the hood or toward an approved grease reservoir. Where horizontal ducts exceed 75 feet (22 860 mm) in length, the slope shall not be less than one unit vertical in 12 units horizontal (8.3-percent slope).

506.3.8 Grease duct cleanouts and other openings. Grease duct systems shall not have openings therein other than those required for proper operation and maintenance of the system. Any portion of such system having sections not provided with access from the duct entry or discharge shall be provided with cleanout openings. Cleanout openings shall be equipped with tight-fitting doors constructed of steel having a thickness not less than that required for the duct. Doors shall be equipped with a substantial method of latching, sufficient to hold the door tightly closed. Doors shall be designed so that they are operable without the use of a tool. Door assemblies, including any frames and gasketing, shall be approved for the purpose, and shall not have fasteners that penetrate the duct. Listed and labeled access door assemblies shall be installed in accordance with the terms of the listing.

506.3.8.1 Personnel entry. Where ductwork is large enough to allow entry of personnel, not less than one approved or listed opening having dimensions not less than 20 inches by 20 inches (508 mm by 508 mm) shall be provided in the horizontal sections, and in the top of vertical risers. Where such entry is provided, the duct and its supports shall be capable of supporting the additional load and the cleanouts specified in Section 506.3.8 are not required.

506.3.9 Grease duct horizontal cleanouts. Cleanouts located on horizontal sections of ducts shall be spaced not more than 20 feet (6096 mm) apart. The cleanouts shall be located on the side of the duct with the opening not less than 1.5 inches (38 mm) above the bottom of the duct, and not less than 1 inch (25 mm) below the top of the duct. The opening minimum dimensions shall be 12 inches (305 mm) on each side. Where the dimensions of the side of the duct prohibit the cleanout installation prescribed herein, the openings shall be on the top of the duct or the bottom of the duct. Where located on the top of the duct, the opening edges shall be a minimum of 1 inch (25 mm) from the edges of the duct. Where located in the bottom of the duct, cleanout openings shall be designed

to provide internal damming around the opening, shall be provided with gasketing to preclude grease leakage, shall provide for drainage of grease down the duct around the dam, and shall be approved for the application. Where the dimensions of the sides, top or bottom of the duct preclude the installation of the prescribed minimum-size cleanout opening, the cleanout shall be located on the duct face that affords the largest opening dimension and shall be installed with the opening edges at the prescribed distances from the duct edges as previously set forth in this section.

506.3.10 Grease duct enclosure. A grease duct serving a Type I hood that penetrates a ceiling, wall or floor shall be enclosed from the point of penetration to the outlet terminal. A duct shall penetrate exterior walls only at locations where unprotected openings are permitted by the *International Building Code*. Ducts shall be enclosed in accordance with the *International Building Code* requirements for shaft construction. The duct enclosure shall be sealed around the duct at the point of penetration and vented to the outside of the building through the use of weather-protected openings. Clearance from the duct to the interior surface of enclosures of combustible construction shall be not less than 18 inches (457 mm). Clearance from the duct to the interior surface of enclosures of noncombustible construction or gypsum wallboard attached to noncombustible structures shall be not less than 6 inches (152 mm). The duct enclosure shall serve a single grease exhaust duct system and shall not contain any other ducts, piping, wiring or systems.

Exceptions:

1. The shaft enclosure provisions of this section shall not be required where a duct penetration is protected with a through-penetration firestop system classified in accordance with ASTM E 814 and having an "F" and "T" rating equal to the fire-resistance rating of the assembly being penetrated and where the surface of the duct is continuously covered on all sides from the point at which the duct penetrates a ceiling, wall or floor to the outlet terminal with a classified and labeled material, system, method of construction or product specifically evaluated for such purpose, in accordance with a nationally recognized standard for such enclosure materials. Exposed duct wrap systems shall be protected where subject to physical damage.

2. A duct enclosure shall not be required for a grease duct that penetrates only a nonfire-resistance-rated roof/ceiling assembly.

506.3.11 Grease duct fire-resistive access opening. Where cleanout openings are located in ducts within a fire-resistance-rated enclosure, access openings shall be provided in the enclosure at each cleanout point. Access openings shall be equipped with tight-fitting sliding or hinged doors that are equal in fire-resistive protection to that of the shaft or enclosure. An approved sign shall be placed on access opening panels with wording as follows: "ACCESS PANEL. DO NOT OBSTRUCT."

506.3.12 Exhaust outlets serving Type I hoods. Exhaust outlets for grease ducts serving Type I hoods shall conform to the requirements of Sections 506.3.12.1 through 506.3.12.3.

506.3.12.1 Termination above the roof. Exhaust outlets that terminate above the roof shall have the discharge opening located not less than 40 inches (1016 mm) above the roof surface.

506.3.12.2 Termination through an exterior wall. Exhaust outlets shall be permitted to terminate through exterior walls where the smoke, grease, gases, vapors, and odors in the discharge from such terminations do not create a public nuisance or a fire hazard. Such terminations shall not be located where protected openings are required by the International Building Code. Other exterior openings shall not be located within 3 feet (914 mm) of such terminations.

506.3.12.3 Termination location. Exhaust outlets shall be located not less than 10 feet (3048 mm) horizontally from parts of the same or contiguous buildings, adjacent property lines and air intake openings into any building and shall be located not less than 10 feet (3048 mm) above the adjoining grade level.

Exception: Exhaust outlets shall terminate not less than 5 feet (1524 mm) from an adjacent building, adjacent property line and air intake openings into a building where air from the exhaust outlet discharges away from such locations.

506.4 Ducts serving Type II hoods. Single or combined Type II exhaust systems for food-processing operations shall be independent of all other exhaust systems. Commercial kitchen exhaust systems serving Type II hoods shall comply with Sections 506.4.1 and 506.4.2.

506.4.1 Type II exhaust outlets. Exhaust outlets for ducts serving Type II hoods shall comply with Sections 401.5 and 401.5.2. Such outlets shall be protected against local weather conditions and shall meet the provisions for exterior wall opening protectives in accordance with the International Building Code.

506.4.2 Ducts. Ducts and plenums serving Type II hoods shall be constructed of rigid metallic materials. Duct construction, installation, bracing and supports shall comply with Chapter 6. Ducts subject to positive pressure and ducts conveying moisture-laden or waste-heat-laden air shall be constructed, joined and sealed in an approved manner.

506.5 Exhaust equipment. Exhaust equipment, including fans and grease reservoirs, shall comply with Section 506.5.1 through 506.5.5 and shall be of an approved design or shall be listed for the application.

506.5.1 Exhaust fans. Exhaust fan housings serving a Type I hood shall be constructed as required for grease ducts in accordance with Section 506.3.1.1.

Exception: Fans listed and labeled in accordance with UL 762.

506.5.1.1 Fan motor. Exhaust fan motors shall be located outside of the exhaust airstream.

506.5.2 Exhaust fan discharge. Exhaust fans shall be positioned so that the discharge will not impinge on the roof,

other equipment or appliances or parts of the structure. A vertical discharge fan shall be manufactured with an approved drain outlet at the lowest point of the housing to permit drainage of grease to an approved grease reservoir.

506.5.3 Exhaust fan mounting. An upblast fan shall be hinged and supplied with a flexible weatherproof electrical cable to permit inspection and cleaning. The ductwork shall extend a minimum of 18 inches (457 mm) above the roof surface.

506.5.4 Clearances. Exhaust equipment serving a Type I hood shall have a clearance to combustible construction of not less than 18 inches (457 mm).

Exception: Factory-built exhaust equipment installed in accordance with Section 304.1 and listed for a lesser clearance.

506.5.5 Termination location. The outlet of exhaust equipment serving Type I hoods, shall be in accordance with Section 506.3.12.3

Exception: The minimum horizontal distance between vertical discharge fans and parapet-type building structures shall be 2 feet (610 mm) provided that such structures are not higher than the top of the fan discharge opening.

SECTION 507
COMMERCIAL KITCHEN HOODS

507.1 General. Commercial kitchen exhaust hoods shall comply with the requirements of this section. Hoods shall be Type I or Type II and shall be designed to capture and confine cooking vapors and residues.

Exceptions:

1. Factory-built commercial exhaust hoods which are tested in accordance with UL 710, listed, labeled and installed in accordance with Section 304.1 shall not be required to comply with Sections 507.4, 507.7, 507.11, 507.12, 507.13, 507.14 and 507.15.

2. Factory-built commercial cooking recirculating systems which are tested in accordance with UL 197, listed, labeled and installed in accordance with Section 304.1 shall not be required to comply with Sections 507.4, 507.5, 507.7, 507.12, 507.13, 507.14 and 507.15.

3. Net exhaust volumes for hoods shall be permitted to be reduced during no-load cooking conditions, where engineered or listed multi-speed or variable-speed controls automatically operate the exhaust system to maintain capture and removal of cooking effluents as required by this section.

507.2 Where required. A Type I or Type II hood shall be installed at or above all commercial cooking appliances in accordance with Sections 507.2.1 and 507.2.2. Where any cooking appliance under a single hood requires a Type I hood, a Type I hood shall be installed. Where a Type II hood is required, a Type I or Type II hood shall be installed.

507.2.1 Type I hoods. Type I hoods shall be installed where cooking appliances produce grease or smoke, such as occurs with griddles, fryers, broilers, ovens, ranges and wok ranges.

507.2.2 Type II hoods. Type II hoods shall be installed where cooking or dishwashing appliances produce heat or steam and do not produce grease or smoke, such as steamers, kettles, pasta cookers and dishwashing machines.

Exceptions:

1. Under-counter-type commercial dishwashing machines.

2. A Type II hood is not required for dishwashers and potwashers that are provided with heat and water vapor exhaust systems that are supplied by the appliance manufacturer and are installed in accordance with the manufacturer's instructions.

507.2.3 Domestic cooking appliances used for commercial purposes. Domestic cooking appliances utilized for commercial purposes shall be provided with Type I or Type II hoods as required for the type of appliances and processes in accordance with Sections 507.2, 507.2.1 and 507.2.2.

507.2.4 Solid fuel. Type I hoods for use over solid fuel-burning cooking appliances shall discharge to an exhaust system that is independent of other exhaust systems.

507.3 Fuel-burning appliances. Where vented fuel-burning appliances are located in the same room or space as the hood, provisions shall be made to prevent the hood system from interfering with normal operation of the appliance vents.

507.4 Type I materials. Type I hoods shall be constructed of steel not less than 0.043 inch (1.09 mm) (No. 18 MSG) in thickness, or stainless steel not less than 0.037 inch (0.94 mm) (No. 20 MSG) in thickness.

507.5 Type II hood materials. Type II hoods shall be constructed of steel not less than 0.030 inch (0.76 mm) (No. 22 Gage) in thickness, stainless steel not less than 0.024 inch (0.61 mm) (No. 24 Gage) in thickness, copper sheets weighing not less than 24 ounces per square foot (7.3 kg/m²), or of other approved material and gage.

507.6 Supports. Type I hoods shall be secured in place by noncombustible supports. All Type I and Type II hood supports shall be adequate for the applied load of the hood, the unsupported ductwork, the effluent loading, and the possible weight of personnel working in or on the hood.

507.7 Hood joints, seams and penetrations. Hood joints, seams and penetrations shall comply with Sections 507.7.1 and 507.7.2.

507.7.1 Type I hoods. External hood joints, seams and penetrations for Type I hoods shall be made with a continuous external liquid-tight weld or braze to the lowest outermost perimeter of the hood. Internal hood joints, seams, penetrations, filter support frames, and other appendages attached inside the hood shall not be required to be welded or brazed but shall be otherwise sealed to be grease tight.

Exceptions:

1. Penetrations shall not be required to be welded or brazed where sealed by devices that are listed for the application.

2. Internal welding or brazing of seams, joints, and penetrations of the hood shall not be prohibited provided that the joint is formed smooth or ground so as to not trap grease, and is readily cleanable.

507.7.2 Type II hoods. Joints, seams and penetrations for Type II hoods shall be constructed as set forth in Chapter 6, shall be sealed on the interior of the hood and shall provide a smooth surface that is readily cleanable and water tight.

507.8 Cleaning and grease gutters. A hood shall be designed to provide for thorough cleaning of the entire hood. Grease gutters shall drain to an approved collection receptacle that is fabricated, designed and installed to allow access for cleaning.

507.9 Clearances for Type I hood. A Type I hood shall be installed with a clearance to combustibles of not less than 18 inches (457 mm).

Exception: Clearance shall not be required from gypsum wallboard attached to noncombustible structures provided that a smooth, cleanable, nonabsorbent and noncombustible material is installed between the hood and the gypsum wallboard over an area extending not less than 18 inches (457 mm) in all directions from the hood.

507.10 Hoods penetrating a ceiling. Type I hoods or portions thereof penetrating a ceiling, wall or furred space shall comply with all the requirements of Section 506.3.10.

507.11 Grease filters. Type I hoods shall be equipped with listed grease filters designed for the specific purpose. Grease-collecting equipment shall be provided with access for cleaning. The lowest edge of a grease filter located above the cooking surface shall be not less than the height specified in Table 507.11.

TABLE 507.11
MINIMUM DISTANCE BETWEEN THE LOWEST EDGE OF A GREASE FILTER AND THE COOKING SURFACE OR THE HEATING SURFACE

TYPE OF COOKING APPLIANCES	HEIGHT ABOVE COOKING SURFACE (feet)
Without exposed flame	0.5
Exposed flame and burners	2
Exposed charcoal and charbroil type	3.5

For SI: 1 foot = 304.8 mm.

507.11.1 Criteria. Filters shall be of such size, type and arrangement as will permit the required quantity of air to pass through such units at rates not exceeding those for which the filter or unit was designed or approved. Filter units shall be installed in frames or holders so as to be readily removable without the use of separate tools, unless designed and installed to be cleaned in place and the system is equipped for such cleaning in place. Removable filter units shall be of a size that will allow them to be cleaned in a dishwashing machine or pot sink. Filter units shall be arranged in place or provided with drip-intercepting devices to prevent grease or other condensate from dripping into food or on food preparation surfaces.

507.11.2 Mounting position. Filters shall be installed at an angle of not less than 45 degrees (0.79 rad) from the horizontal and shall be equipped with a drip tray beneath the lower edge of the filters.

507.12 Canopy size and location. The inside lower edge of canopy-type commercial cooking hoods shall overhang or extend a horizontal distance of not less than 6 inches (152 mm) beyond the edge of the cooking surface, on all open sides. The vertical distance between the front lower lip of the hood and the cooking surface shall not exceed 4 feet (1219 mm).

Exception: The hood shall be permitted to be flush with the outer edge of the cooking surface where the hood is closed to the appliance side by a noncombustible wall or panel.

507.13 Capacity of hoods. Commercial food service hoods shall exhaust a minimum net quantity of air determined in accordance with this section and Sections 507.13.1 through 507.13.4. The net quantity of exhaust air shall be calculated by subtracting any airflow supplied directly to a hood cavity from the total exhaust flow rate of a hood. Where any combination of extra-heavy-duty, heavy-duty, medium-duty, and light-duty cooking appliances are utilized under a single hood, the highest exhaust rate required by this section shall be used for the entire hood.

507.13.1 Extra-heavy-duty cooking appliances. The minimum net airflow for Type I hoods used for extra-heavy-duty cooking appliances shall be determined as follows:

Type of Hood	CFM per linear foot of hood
Wall-mounted canopy	550
Single island canopy	700
Double island canopy (per side)	550
Backshelf/pass-over	Not allowed
Eyebrow	Not allowed

For SI: 1 cfm per linear foot = 1.55 L/s per linear meter.

507.13.2 Heavy-duty cooking appliances. The minimum net airflow for Type I hoods used for heavy-duty cooking appliances shall be determined as follows:

Type of Hood	CFM per linear foot of hood
Wall-mounted canopy	400
Single island canopy	600
Double island canopy (per side)	400
Backshelf/pass-over	400
Eyebrow	Not allowed

For SI: 1 cfm per linear foot = 1.55 L/s per linear meter.

507.13.3 Medium-duty cooking appliances. The minimum net airflow for Type I hoods used for medium-duty cooking appliances shall be determined as follows:

Type of Hood	CFM per linear foot of hood
Wall-mounted canopy	300
Single island canopy	500
Double island canopy (per side)	300
Backshelf/pass-over	300
Eyebrow	250

For SI: 1 cfm per linear foot = 1.55 L/s per linear meter.

507.13.4 Light-duty cooking appliances. The minimum net airflow for Type I hoods used for light duty cooking appliances and food service preparation and cooking operations approved for use under a Type II hood shall be determined as follows:

Type of Hood	CFM per linear foot of hood
Wall-mounted canopy	200
Single island canopy	400
Double island canopy (per side)	250
Backshelf/pass-over	250
Eyebrow	250

For SI: 1 cfm per linear foot = 1.55 L/s per linear meter.

507.14 Noncanopy size and location. Noncanopy-type hoods shall be located a maximum of 3 feet (914 mm) above the cooking surface. The edge of the hood shall be set back a maximum of 1 foot (305 mm) from the edge of the cooking surface.

507.15 Exhaust outlets. Exhaust outlets located within the hood shall be located so as to optimize the capture of particulate matter. Each outlet shall serve not more than a 12-foot (3658 mm) section of hood.

507.16 Performance test. A performance test shall be conducted upon completion and before final approval of the installation of a ventilation system serving commercial cooking appliances. The test shall verify the rate of exhaust airflow required by Section 507.13, makeup airflow required by Section 508, and proper operation as specified in this chapter. The permit holder shall furnish the necessary test equipment and devices required to perform the tests.

507.16.1 Capture and containment test. The permit holder shall verify capture and containment performance of the exhaust system. This field test shall be conducted with all appliances under the hood at operating temperatures. Capture and containment shall be verified visually by observing smoke or steam produced by actual or simulated cooking, such as with smoke candles, smoke puffers, etc.

SECTION 508
COMMERCIAL KITCHEN MAKEUP AIR

508.1 Makeup air. Makeup air shall be supplied during the operation of commercial kitchen exhaust systems that are provided for commercial cooking appliances. The amount of makeup air supplied shall be approximately equal to the amount of exhaust air. The makeup air shall not reduce the effectiveness of the exhaust system. Makeup air shall be provided by gravity or mechanical means or both. For mechanical makeup air systems, the exhaust and makeup air systems shall be electrically interlocked to insure that makeup air is provided whenever the exhaust system is in operation. Makeup air intake opening locations shall comply with Sections 401.5 and 401.5.1.

508.1.1 Makeup air temperature. The temperature differential between makeup air and the air in the conditioned space shall not exceed 10°F (6°C).

Exceptions:

1. Makeup air that is part of the air-conditioning system.

2. Makeup air that does not decrease the comfort conditions of the occupied space.

508.2 Compensating hoods. Manufacturers of compensating hoods shall provide a label indicating minimum exhaust flow and/or maximum makeup airflow that provides capture and containment of the exhaust effluent.

SECTION 509
FIRE SUPPRESSION SYSTEMS

509.1 Where required. Commercial food heat-processing appliances required by Section 507.2.1 to have a Type I hood shall be provided with an approved automatic fire suppression system complying with the *International Building Code* and the *International Fire Code*.

SECTION 510
HAZARDOUS EXHAUST SYSTEMS

510.1 General. This section shall govern the design and construction of duct systems for hazardous exhaust and shall determine where such systems are required. Hazardous exhaust systems are systems designed to capture and control hazardous emissions generated from product handling or processes, and convey those emissions to the outdoors. Hazardous emissions include flammable vapors, gases, fumes, mists or dusts, and volatile or airborne materials posing a health hazard, such as toxic or corrosive materials. For the purposes of this section, the health-hazard rating of materials shall be as specified in NFPA 704.

510.2 Where required. A hazardous exhaust system shall be required wherever operations involving the handling or processing of hazardous materials, in the absence of such exhaust systems and under normal operating conditions, have the potential to create one of the following conditions:

1. A flammable vapor, gas, fume, mist or dust is present in concentrations exceeding 25 percent of the lower flammability limit of the substance for the expected room temperature.

2. A vapor, gas, fume, mist or dust with a health-hazard rating of 4 is present in any concentration.

3. A vapor, gas, fume, mist or dust with a health-hazard rating of 1, 2 or 3 is present in concentrations exceeding 1 percent of the median lethal concentration of the substance for acute inhalation toxicity.

[F] 510.2.1 Lumber yards and woodworking facilities. Equipment or machinery located inside buildings at lumber yards and woodworking facilities which generates or emits combustible dust shall be provided with an approved dust-collection and exhaust system installed in conformance with this section and the *International Fire Code*. Equipment and systems that are used to collect, process or convey combustible dusts shall be provided with an approved explosion-control system.

[F] 510.2.2 Combustible fibers. Equipment or machinery within a building which generates or emits combustible fibers shall be provided with an approved dust-collecting and

exhaust system. Such systems shall comply with this code and the *International Fire Code*.

510.3 Design and operation. The design and operation of the exhaust system shall be such that flammable contaminants are diluted in noncontaminated air to maintain concentrations in the exhaust flow below 25 percent of the contaminant's lower flammability limit.

510.4 Independent system. Hazardous exhaust systems shall be independent of other types of exhaust systems. Incompatible materials, as defined in the *International Fire Code*, shall not be exhausted through the same hazardous exhaust system. Hazardous exhaust systems shall not share common shafts with other duct systems, except where such systems are hazardous exhaust systems originating in the same fire area.

Contaminated air shall not be recirculated to occupied areas unless the contaminants have been removed. Air contaminated with explosive or flammable vapors, fumes or dusts; flammable or toxic gases; or radioactive material shall not be recirculated.

510.5 Design. Systems for removal of vapors, gases and smoke shall be designed by the constant velocity or equal friction methods. Systems conveying particulate matter shall be designed employing the constant velocity method.

510.5.1 Balancing. Systems conveying explosive or radioactive materials shall be prebalanced by duct sizing. Other systems shall be balanced by duct sizing with balancing devices, such as dampers. Dampers provided to balance air-flow shall be provided with securely fixed minimum-position blocking devices to prevent restricting flow below the required volume or velocity.

510.5.2 Emission control. The design of the system shall be such that the emissions are confined to the area in which they are generated by air currents, hoods or enclosures and shall be exhausted by a duct system to a safe location or treated by removing contaminants.

510.5.3 Hoods required. Hoods or enclosures shall be used where contaminants originate in a limited area of a space. The design of the hood or enclosure shall be such that air currents created by the exhaust systems will capture the contaminants and transport them directly to the exhaust duct.

510.5.4 Contaminant capture and dilution. The velocity and circulation of air in work areas shall be such that contaminants are captured by an airstream at the area where the emissions are generated and conveyed into a product-conveying duct system. Contaminated air from work areas where hazardous contaminants are generated shall be diluted below the thresholds specified in Section 510.2 with air that does not contain other hazardous contaminants.

510.5.5 Makeup air. Makeup air shall be provided at a rate approximately equal to the rate that air is exhausted by the hazardous exhaust system. Makeup-air intakes shall be located so as to avoid recirculation of contaminated air.

510.5.6 Clearances. The minimum clearance between hoods and combustible construction shall be the clearance required by the duct system.

510.5.7 Ducts. Hazardous exhaust duct systems shall extend directly to the exterior of the building and shall not extend into or through ducts and plenums.

510.6 Penetrations. Penetrations of structural elements by a hazardous exhaust system shall conform to Sections 510.6.1 through 510.6.3.

Exception: Duct penetrations within H-5 occupancies as allowed by the *International Building Code*.

510.6.1 Floors. Hazardous exhaust systems that penetrate a floor/ceiling assembly shall be enclosed in a fire-resistance-rated shaft constructed in accordance with the *International Building Code*.

510.6.2 Wall assemblies. Hazardous exhaust duct systems that penetrate fire-resistance-rated wall assemblies shall be enclosed in fire-resistance-rated construction from the point of penetration to the outlet terminal, except where the interior of the duct is equipped with an approved automatic fire suppression system. Ducts shall be enclosed in accordance with the *International Building Code* requirements for shaft construction and such enclosure shall have a minimum fire-resistance-rating of not less than the highest fire-resistance-rated wall assembly penetrated.

510.6.3 Fire walls. Ducts shall not penetrate a fire wall.

510.7 Suppression required. Ducts shall be protected with an approved automatic fire suppression system installed in accordance with the *International Building Code*.

Exceptions:

1. An approved automatic fire suppression system shall not be required in ducts conveying materials, fumes, mists and vapors that are nonflammable and noncombustible under all conditions and at any concentrations.

2. An approved automatic fire suppression system shall not be required in ducts where the largest cross-sectional diameter of the duct is less than 10 inches (254 mm).

510.8 Duct construction. Ducts utilized to convey hazardous exhaust shall be constructed of approved G90 galvanized sheet steel, with a minimum nominal thickness as specified in Table 510.8.

Nonmetallic ducts utilized in systems exhausting nonflammable corrosive fumes or vapors shall be listed and labeled. Nonmetallic duct shall have a flame spread index of 25 or less and a smoke-developed index of 50 or less, when tested in accordance with ASTM E 84. Ducts shall be approved for installation in such an exhaust system.

Where the products being exhausted are detrimental to the duct material, the ducts shall be constructed of alternative materials that are compatible with the exhaust.

510.8.1 Duct joints. Ducts shall be made tight with lap joints having a minimum lap of 1 inch (25 mm).

510.8.2 Clearance to combustibles. Ducts shall have a clearance to combustibles in accordance with Table 510.8.2. Exhaust gases having temperatures in excess of 600°F (316°C) shall be exhausted to a chimney in accordance with Section 511.2.

TABLE 510.8
MINIMUM DUCT THICKNESS

DIAMETER OF DUCT OR MAXIMUM SIDE DIMENSION	MINIMUM NOMINAL THICKNESS		
	Nonabrasive materials	Nonabrasive/ Abrasive materials	Abrasive materials
0-8 inches	0.028 inch (No. 24 Gage)	0.034 inch (No. 22 Gage)	0.040 inch (No. 20 Gage)
9-18 inches	0.034 inch (No. 22 Gage)	0.040 inch (No. 20 Gage)	0.052 inch (No. 18 Gage)
19-30 inches	0.040 inch (No. 20 Gage)	0.052 inch (No. 18 Gage)	0.064 inch (No. 16 Gage)
Over 30 inches	0.052 inch (No. 18 Gage)	0.064 inch (No. 16 Gage)	0.079 inch (No. 14 Gage)

For SI:1 inch = 25.4 mm.

510.8.3 Explosion relief. Systems exhausting potentially explosive mixtures shall be protected with an approved explosion relief system or by an approved explosion prevention system designed and installed in accordance with NFPA 69. An explosion relief system shall be designed to minimize the structural and mechanical damage resulting from an explosion or deflagration within the exhaust system. An explosion prevention system shall be designed to prevent an explosion or deflagration from occurring.

TABLE 510.8.2
CLEARANCE TO COMBUSTIBLES

TYPE OF EXHAUST OR TEMPERATURE OF EXHAUST (°F)	CLEARANCE TO COMBUSTIBLES (inches)
Less than 100	1
100-600	12
Flammable vapors	6

For SI: 1 inch = 25.4 mm, EC = [(EF)- 32]/1.8.

510.9 Supports. Ducts shall be supported at intervals not exceeding 10 feet (3048 mm). Supports shall be constructed of noncombustible material.

SECTION 511
DUST, STOCK AND REFUSE CONVEYING SYSTEMS

511.1 Dust, stock and refuse conveying systems. Dust, stock and refuse conveying systems shall comply with the provisions of Section 510 and Sections 511.1.1 through 511.2.

511.1.1 Collectors and separators. Cyclone collectors and separators and associated supports shall be constructed of noncombustible materials and shall be located on the exterior of the building or structure. A collector or separator shall not be located nearer than 10 feet (3048 mm) to combustible construction or to an unprotected wall or floor opening, unless the collector is provided with a metal vent pipe that extends above the highest part of any roof within a distance of 30 feet (9144 mm).

511.1.2 Discharge pipe. Discharge piping shall conform to the requirements for ducts, including clearances required for high-heat appliances, as contained in this code. A delivery pipe from a cyclone collector shall not convey refuse directly into the firebox of a boiler, furnace, dutch oven, refuse burner, incinerator or other appliance.

511.1.3 Conveying system exhaust discharge. An exhaust system shall discharge to the outside of the building either directly by flue, or indirectly through the separator, bin or vault into which the system discharges.

511.1.4 Spark protection. The outlet of an open-air exhaust terminal shall be protected with an approved metal or other noncombustible screen to prevent the entry of sparks.

511.1.5 Explosion relief vents. A safety or explosion relief vent shall be provided on all systems that convey combustible refuse or stock of an explosive nature, in accordance with the requirements of the *International Building Code.*

511.1.5.1 Screens. Where a screen is installed in a safety relief vent, the screen shall be attached so as to permit ready release under the explosion pressure.

511.1.5.2 Hoods. The relief vent shall be provided with an approved noncombustible cowl or hood, or with a counterbalanced relief valve or cover arranged to prevent the escape of hazardous materials, gases or liquids.

511.2 Exhaust outlets. Outlets for exhaust that exceed 600°F (315°C) shall be designed as a chimney in accordance with Table 511.2.

The termination point for exhaust ducts discharging to the atmosphere shall not be less than the following:

1. Ducts conveying explosive or flammable vapors, fumes or dusts: 30 feet (9144 mm) from property line; 10 feet (3048 mm) from openings into the building; 6 feet (1829 mm) from exterior walls or roofs; 30 feet (9144 mm) from combustible walls or openings into the building which are in the direction of the exhaust discharge; and 10 feet (3048 mm) above adjoining grade.

2. Other product-conveying outlets: 10 feet (3048 mm) from property line; 3 feet (914 mm) from exterior wall or roof; 10 feet (3048 mm) from openings into the building; and 10 feet (3048 mm) above adjoining grade.

3. Environmental air duct exhaust: 3 feet (914 mm) from property line; and 3 feet (914 mm) from openings into the building.

SECTION 512
SUBSLAB SOIL EXHAUST SYSTEMS

512.1 General. When a subslab soil exhaust system is provided, the duct shall conform to the requirements of this section.

512.2 Materials. Subslab soil exhaust system duct material shall be air duct material listed and labeled to the requirements of UL 181 for Class 0 air ducts, or any of the following piping materials that comply with the *International Plumbing Code* as building sanitary drainage and vent pipe: cast iron; galvanized steel; brass or copper pipe; copper tube of a weight not less than that of copper drainage tube, Type DWV; and plastic piping.

TABLE 511.2
CONSTRUCTION, CLEARANCE AND TERMINATION REQUIREMENTS FOR SINGLE-WALL METAL CHIMNEYS

CHIMNEYS SERVING	MINIMUM THICKNESS		TERMINATION				CLEARANCE			
	Walls	Lining	Above roof opening (feet)	Above any part of building within (feet)			Combustible construction (inches)		Noncombustible construction	
				10	25	50	Interior inst.	Exterior inst.	Interior inst.	Exterior inst.
Low-heat appliances (1,000°F normal operation)	0.127" (No. 10 MSG)	None	3	2	—	—	18	6	Up to 18" diameter, 2" Over 18" diameter, 4"	
Medium-heat appliances (2,000°F maximum)[b]	0.127" (No. 10 MSG)	Up to 18" dia.—2¹/₂" Over 18"-4¹/₂" On 4¹/₂" bed	10	—	10	—	36	24		
High-heat appliances (Over 2,000°F)[a]	0.127" (No. 10 MSG)	4¹/₂" laid on 4¹/₂" bed	20	—	—	20	See Note c			

For SI: 1 inch = 25.4 mm, 1 foot = 304.8 mm, °C = [(°F)-32]/1.8.

a. Lining shall extend from bottom to top of outlet.

b. Lining shall extend from 24 inches below connector to 24 feet above.

c. Clearance shall be as specified by the design engineer and shall have sufficient clearance from buildings and structures to avoid overheating combustible materials (maximum 160°F).

512.3 Grade. Exhaust system ducts shall not be trapped and shall have a minimum slope of one-eighth unit vertical in 12 units horizontal (1-percent slope).

512.4 Termination. Subslab soil exhaust system ducts shall extend through the roof and terminate at least 6 inches (152 mm) above the roof and at least 10 feet (3048 mm) from any operable openings or air intake.

512.5 Identification. Subslab soil exhaust ducts shall be permanently identified within each floor level by means of a tag, stencil or other approved marking.

SECTION 513
SMOKE CONTROL SYSTEMS

[B] 513.1 Scope and purpose. This section applies to mechanical and passive smoke control systems that are required by the *International Building Code*. The purpose of this section is to establish minimum requirements for the design, installation and acceptance testing of smoke control systems that are intended to provide a tenable environment for the evacuation or relocation of occupants. These provisions are not intended for the preservation of contents, the timely restoration of operations, or for assistance in fire suppression or overhaul activities. Smoke control systems regulated by this section serve a different purpose than the smoke- and heat-venting provisions found in Section 910 of the *International Building Code.*

[B] 513.2 General design requirements. Buildings, structures, or parts thereof required by this code to have a smoke control system or systems shall have such systems designed in accordance with the applicable requirements of Section 909 of the *International Building Code* and the generally accepted and well-established principles of engineering relevant to the design. The construction documents shall include sufficient information and detail to describe adequately the elements of the design necessary for the proper implementation of the smoke control systems. These documents shall be accompanied with sufficient information and analysis to demonstrate compliance with these provisions

[B] 513.3 Special inspection and test requirements. In addition to the ordinary inspection and test requirements which buildings, structures and parts thereof are required to undergo, smoke control systems subject to the provisions of Section 909 of the *International Building Code* shall undergo special inspections and tests sufficient to verify the proper commissioning of the smoke control design in its final installed condition. The design submission accompanying the construction documents shall clearly detail procedures and methods to be used and the items subject to such inspections and tests. Such commissioning shall be in accordance with generally accepted engineering practice and, where possible, based on published standards for the particular testing involved. The special inspections and tests required by this section shall be conducted under the same terms as found in Section 1704 of the *International Building Code.*

[B] 513.4 Analysis. A rational analysis supporting the types of smoke control systems to be employed, their methods of operation, the systems supporting them, and the methods of construction to be utilized shall accompany the submitted construction documents and shall include, but not be limited to, the items indicated in Sections 513.4.1 through 513.4.6.

[B] 513.4.1 Stack effect. The system shall be designed such that the maximum probable normal or reverse stack effects will not adversely interfere with the system's capabilities. In determining the maximum probable stack effects, altitude, elevation, weather history and interior temperatures shall be used.

[B] 513.4.2 Temperature effect of fire. Buoyancy and expansion caused by the design fire in accordance with Section 513.9 shall be analyzed. The system shall be designed such that these effects do not adversely interfere with its capabilities.

[B] 513.4.3 Wind effect. The design shall consider the adverse effects of wind. Such consideration shall be consistent with the wind-loading provisions of the *International Building Code.*

[B] 513.4.4 HVAC systems. The design shall consider the effects of the heating, ventilating and air-conditioning (HVAC) systems on both smoke and fire transport. The analysis shall include all permutations of systems' status. The design shall consider the effects of fire on the HVAC systems.

[B] 513.4.5 Climate. The design shall consider the effects of low temperatures on systems, property and occupants. Air inlets and exhausts shall be located so as to prevent snow or ice blockage.

[B] 513.4.6 Duration of operation. All portions of active or passive smoke control systems shall be capable of continued operation after detection of the fire event for not less than 20 minutes.

[B] 513.5 Smoke barrier construction. Smoke barriers shall comply with the *International Building Code.* Smoke barriers shall be constructed and sealed to limit leakage areas exclusive of protected openings. The maximum allowable leakage area shall be the aggregate area calculated using the following leakage area ratios:

1. Walls: $A/A_w = 0.00100$

2. Exit enclosures: $A/A_w = 0.00035$

3. All other shafts: $A/A_w = 0.00150$

4. Floors and roofs: $A/A_F = 0.00050$

where:

A = Total leakage area, square feet (m^2).

A_F = Unit floor or roof area of barrier, square feet (m^2).

A_w = Unit wall area of barrier, square feet (m^2).

The leakage area ratios shown do not include openings due to doors, operable windows or similar gaps. These shall be included in calculating the total leakage area.

[B] 513.5.1 Leakage area. Total leakage area of the barrier is the product of the smoke barrier gross area times the allowable leakage area ratio. Compliance shall be determined by achieving the minimum air pressure difference across the barrier with the system in the smoke control mode for mechanical smoke control systems. Passive smoke control systems tested using other approved means such as door fan testing shall be as approved by the code official.

[B] 513.5.2 Opening protection. Openings in smoke barriers shall be protected by automatic-closing devices actuated by the required controls for the mechanical smoke control system. Door openings shall be protected by door assemblies complying with the requirements of the *International Building Code* for doors in smoke barriers.

Exceptions:

1. Passive smoke control systems with automatic-closing devices actuated by spot-type smoke detectors listed for releasing service installed in accordance with the *International Building Code.*

2. Fixed openings between smoke zones which are protected utilizing the airflow method.

3. In Group I-2 where such doors are installed across corridors, a pair of opposite-swinging doors without a center mullion shall be installed having vision panels with approved fire-rated glazing materials in approved fire-rated frames, the area of which shall not exceed that tested. The doors shall be close-fitting within operational tolerances, and shall not have undercuts, louvers or grilles. The doors shall have head and jamb stops, astragals or rabbets at meeting edges and automatic-closing devices. Positive latching devices are not required.

4. Group I-3.

5. Openings between smoke zones with clear ceiling heights of 14 feet (4267 mm) or greater and bank down capacity of greater than 20 minutes as determined by the design fire size.

[B] 513.5.2.1 Ducts and air transfer openings. Ducts and air transfer openings are required to be protected with a minimum Class II, 250°F (121°C) smoke damper complying with the *International Building Code.*

[B] 513.6 Pressurization method. The primary mechanical means of controlling smoke shall be by pressure differences across smoke barriers. Maintenance of a tenable environment is not required in the smoke control zone of fire origin.

[B] 513.6.1 Minimum pressure difference. The minimum pressure difference across a smoke barrier shall be 0.05-inch water gage (12.4 Pa) in fully sprinklered buildings.

In buildings permitted to be other than fully sprinklered, the smoke control system shall be designed to achieve pressure differences at least two times the maximum calculated pressure difference produced by the design fire.

[B] 513.6.2 Maximum pressure difference. The maximum air pressure difference across a smoke barrier shall be determined by required door-opening or closing forces. The actual force required to open exit doors when the system is in the smoke control mode shall be in accordance with the *International Building Code.* Opening and closing forces for other doors shall be determined by standard engineering methods for the resolution of forces and reactions. The calculated force to set a side-hinged, swinging door in motion shall be determined by:

$$F = F_{dc} + K(WA\Delta P)/2(W\text{-}d) \qquad \textbf{(Equation 5-2)}$$

where:

A = Door area, square feet (m²).

d = Distance from door handle to latch edge of door, feet (m).

F = Total door opening force, pounds (N).

F_{dc} = Force required to overcome closing device, pounds (N).

K = Coefficient 5.2 (1.0).

W = Door width, feet (m).

$\triangle P$ = Design pressure difference, inches (Pa) water gage.

[B] 513.7 Airflow design method. When approved by the code official, smoke migration through openings fixed in a permanently open position, which are located between smoke control zones by the use of the airflow method, shall be permitted. The design airflows shall be in accordance with this section. Air-flow shall be directed to limit smoke migration from the fire zone. The geometry of openings shall be considered to prevent flow reversal from turbulent effects.

[B] 513.7.1 Velocity. The minimum average velocity through a fixed opening shall not be less than:

$$v = 217.2\ [h\ (T_f - T_o)/(T_f + 460)]^{1/2} \qquad \textbf{(Equation 5-3)}$$

For SI: $v = 119.9\ [h\ (T_f - T_o)/T_f]^{1/2}$

where:

H = Height of opening, feet (m).

T_f = Temperature of smoke, °F (K).

T_o = Temperature of ambient air, °F (K).

v = Air velocity, feet per minute (m/minute).

[B] 513.7.2 Prohibited conditions. This method shall not be employed where either the quantity of air or the velocity of the airflow will adversely affect other portions of the smoke control system, unduly intensify the fire, disrupt plume dynamics or interfere with exiting. In no case shall airflow toward the fire exceed 200 feet per minute (1.02 m/s). Where the formula in Section 513.7.1 requires airflow to exceed this limit, the airflow method shall not be used.

[B] 513.8 Exhaust method. When approved by the code official, mechanical smoke control for large enclosed volumes, such as in atria or malls, shall be permitted to utilize the exhaust method. The design exhaust volumes shall be in accordance with this section.

[B] 513.8.1 Exhaust rate. The height of the lowest horizontal surface of the accumulating smoke layer shall be maintained at least 10 feet (3048 mm) above any walking surface which forms a portion of a required egress system within the smoke zone. The required exhaust rate for the zone shall be the largest of the calculated plume mass flow rates for the possible plume configurations. Provisions shall be made for natural or mechanical supply of outside air from outside or adjacent smoke zones to make up for the air exhausted. Makeup airflow rates, when measured at the potential fire location, shall not exceed 200 feet per minute (1.02 m/s) toward the fire. The temperature of the makeup air shall be such that it does not expose temperature-sensitive fire protection systems beyond their limits.

[B] 513.8.2 Axisymmetric plumes. The plume mass flow rate (m_p), in pounds per second (kg/s), shall be determined by placing the design fire center on the axis of the space being analyzed. The limiting flame height shall be determined by:

$$z_l = 0.533Q_c^{2/5} \qquad \text{(Equation 5-4)}$$

For SI: $z_l = 0.166Q_c^{2/5}$

where:

M_p = Plume mass flow rate, pounds per second (kg/s).

Q = Total heat output.

Q_c = Convective heat output, British thermal units per second (kW).

(The value of Q_c shall not be taken as less than $0.70Q$).

z = Height from top of fuel surface to bottom of smoke layer, feet (m).

z_l = Limiting flame height, feet (m). The z_l value must be greater than the fuel equivalent diameter (see Section 513.9).

for $z > z_l$

$$m_p = 0.022Q_c^{1/3}z^{5/3} + 0.0042Q_c$$

For SI: $m_p = 0.071Q_c^{1/3}z^{5/3} + 0.0018Q_c$

for $z = z_l$

$$M_p = 0.011 Q_c$$

For SI: $m_p = 0.035Q_c$

for $z < z_l$

$$M_p = 0.0208Q_c^{3/5}z$$

For SI: $m_p = 0.032Q_c^{3/5}z$

To convert m_p from pounds per second of mass flow to a volumetric rate, the following formula shall be used:

$$V = 60m_p/p \qquad \text{(Equation 5-5)}$$

where:

V = Volumetric flow rate, cubic feet per minute (m³/s).

R = Density of air at the temperature of the smoke layer, pounds per cubic feet (T: in °F)[kg/m³ (T: in °C)].

[B] 513.8.3 Balcony spill plumes. The plume mass flow rate (m_p) for spill plumes shall be determined using the geometrically probable width based on architectural elements and projections in the following formula:

$$M_p = 0.124(QW^2)^{1/3}(z_b + 0.25H) \qquad \text{(Equation 5-6)}$$

For SI: $m_p = 0.36(QW^2)^{1/3}(z_b + 0.25H)$

where:

H = Height above fire to underside of balcony, feet (m).

M_p = Plume mass flow rate, pounds per second (kg/s).

Q = Total heat output.

W = Plume width at point of spill, feet (m).

z_b = Height from balcony, feet (m).

[B] 513.8.4 Window plumes. The plume mass flow rate (m_p) shall be determined from:

$$m_p = 0.077(A_wH_w^{1/2})^{1/3}(z_w + a)^{5/3} + 0.18A_wH_w^{1/2} \qquad \text{(Equation 5-7)}$$

For SI: $m_p = 0.68(A_wH_w^{1/2})^{1/3}(z_w + a)^{5/3} + 1.5A_wH_w^{1/2}$

where:

A_w = Area of the opening, square feet (m²).

H_w = Height of the opening, feet (m).

M_p = Plume mass flow rate, pounds per second (kg/s).

z_w = Height from the top of the window or opening to the bottom of the smoke layer, feet (m).

$$a = 2.4A_w^{2/5}H_w^{1/5} - 2.1H_w$$

[B] 513.8.5 Plume contact with walls. When a plume contacts one or more of the surrounding walls, the mass flow rate shall be adjusted for the reduced entrainment resulting from the contact provided that the contact remains constant. Use of this provision requires calculation of the plume diameter, that shall be calculated by:

$$d = 0.48 [(T_c + 460)/(T_a + 460)]^{1/2}z \qquad \text{(Equation 5-8)}$$

For SI: $d = 0.48 (T_c/T_a)^{1/2}z$

where:

d = Plume diameter, feet (m).

T_a = Ambient air temperature, °F (°K).

T_c = Plume centerline temperature, °F (°K).

$\quad = 0.6 (T_a + 460) Q_c^{2/3} Z^{-5/3} + T_a$

z = Height at which T_c is determined, feet (m).

For SI: $T_c = 0.08 T_a Q_c^{2/3} Z^{-5/3} + T_a$

[B] 513.9 Design fire. The design fire shall be based on a Q of not less than 5,000 Btu per second (5275kW) unless a rational analysis is performed by the registered design professional and approved by the code official. The design fire shall be based on the analysis in accordance with Section 513.4 and this section.

[B] 513.9.1 Factors considered. The engineering analysis shall include the characteristics of the fuel, fuel load, effects included by the fire, and whether the fire is likely to be steady or unsteady.

[B] 513.9.2 Separation distance. Determination of the design fire shall include consideration of the type of fuel, fuel spacing and configuration. The ratio of the separation distance to the fuel equivalent radius shall not be less than 4. The fuel equivalent radius shall be the radius of a circle of equal area to floor area of the fuel package. The design fire shall be increased if other combustibles are within the separation distance as determined by:

$$R = [Q/(12\pi q'')]^{1/2} \qquad \text{(Equation 5-9)}$$

where:

Q'' = Incident radiant heat flux required for nonpiloted ignition, Btu/ft² • s (W/m²).

Q = Heat release from fire, Btu/s (kW).

R = Separation distance from target to center of fuel package, feet (m).

[B] 513.9.3 Heat-release assumptions. The analysis shall make use of the best available data from approved sources and shall not be based on excessively stringent limitations of combustible material.

[B] 513.9.4 Sprinkler effectiveness assumptions. A documented engineering analysis shall be provided for conditions that assume fire growth is halted at the time of sprinkler activation.

[B] 513.10 Equipment. Equipment such as, but not limited to, fans, ducts, automatic dampers and balance dampers shall be suitable for their intended use, suitable for the probable exposure temperatures that the rational analysis indicates, and as approved by the code official.

[B] 513.10.1 Exhaust fans. Components of exhaust fans shall be rated and certified by the manufacturer for the probable temperature rise to which the components will be exposed. This temperature rise shall be computed by:

$$T_s = (Q_c/mc) + (T_a) \qquad \textbf{(Equation 5-10)}$$

where:

C = Specific heat of smoke at smoke-layer temperature, Btu/lb°F (kJ/kg × K).

m = Exhaust rate, pounds per second (kg/s).

Q_c = Convective heat output of fire, Btu/s (kW).

T_a = Ambient temperature, °F (K).

T_s = Smoke temperature, °F (K).

Exception: Reduced T_s as calculated based on the assurance of adequate dilution air.

[B] 513.10.2 Ducts. Duct materials and joints shall be capable of withstanding the probable temperatures and pressures to which they are exposed as determined in accordance with Section 513.10.1. Ducts shall be constructed and supported in accordance with Chapter 6. Ducts shall be leak tested to 1.5 times the maximum design pressure in accordance with nationally accepted practices. Measured leakage shall not exceed 5 percent of design flow. Results of such testing shall be a part of the documentation procedure. Ducts shall be supported directly from fire-resistance-rated structural elements of the building by substantial, noncombustible supports.

Exception: Flexible connections, for the purpose of vibration isolation, that are constructed of approved fire-resistance-rated materials.

[B] 513.10.3 Equipment, inlets and outlets. Equipment shall be located so as to not expose uninvolved portions of the building to an additional fire hazard. Outdoor air inlets shall be located so as to minimize the potential for introducing smoke or flame into the building. Exhaust outlets shall be so located as to minimize reintroduction of smoke into the building and to limit exposure of the building or adjacent buildings to an additional fire hazard.

[B] 513.10.4 Automatic dampers. Automatic dampers, regardless of the purpose for which they are installed within the smoke control system, shall be listed and conform to the requirements of approved recognized standards.

[B] 513.10.5 Fans. In addition to other requirements, belt-driven fans shall have 1.5 times the number of belts required for the design duty with the minimum number of belts being two. Fans shall be selected for stable performance based on normal temperature and, where applicable, elevated temperature. Calculations and manufacturer's fan curves shall be part of the documentation procedures. Fans shall be supported and restrained by noncombustible devices in accordance with the structural design requirements of the *International Building Code*. Motors driving fans shall not be operating beyond their nameplate horsepower (kilowatts) as determined from measurement of actual current draw. Motors driving fans shall have a minimum service factor of 1.15.

[B] 513.11 Power systems. The smoke control system shall be supplied with two sources of power. Primary power shall be the normal building power systems. Secondary power shall be from an approved standby source complying with the *ICC Electrical Code*. The standby power source and its transfer switches shall be in a separate room from the normal power transformers and switch gear and shall be enclosed in a room constructed of not less than 1-hour fire-resistance-rated fire barriers, ventilated directly to and from the exterior. Power distribution from the two sources shall be by independent routes. Transfer to full standby power shall be automatic and within 60 seconds of failure of the primary power. The systems shall comply with the *ICC Electrical Code*.

[B] 513.11.1 Power sources and power surges. Elements of the smoke management system relying on volatile memories or the like shall be supplied with uninterruptible power sources of sufficient duration to span 15-minute primary power interruption. Elements of the smoke management system susceptible to power surges shall be suitably protected by conditioners, suppressors or other approved means.

[B] 513.12 Detection and control systems. Fire detection systems providing control input or output signals to mechanical smoke control systems or elements thereof shall comply with the requirements of Chapter 9 of the *International Building Code* and NFPA 72. Such systems shall be equipped with a control unit complying with UL 864 and listed as smoke control equipment.

Control systems for mechanical smoke control systems shall include provisions for verification. Verification shall include positive confirmation of actuation, testing, manual override, the presence of power downstream of all disconnects and, through a preprogrammed weekly test sequence report, abnormal conditions audibly, visually and by printed report.

[B] 513.12.1 Wiring. In addition to meeting the requirements of the *ICC Electrical Code*, all wiring, regardless of voltage, shall be fully enclosed within continuous raceways.

[F] 513.12.2 Activation. Smoke control systems shall be activated in accordance with the *International Building Code*.

[F] 513.12.3 Automatic control. Where completely automatic control is required or used, the automatic control sequences shall be initiated from an appropriately zoned automatic sprinkler system complying with Section

903.3.1.1 of the *International Fire Code* or from manual controls that are readily accessible to the fire department, and any smoke detectors required by engineering analysis.

[B] 513.13 Control-air tubing. Control-air tubing shall be of sufficient size to meet the required response times. Tubing shall be flushed clean and dry prior to final connections. Tubing shall be adequately supported and protected from damage. Tubing passing through concrete or masonry shall be sleeved and protected from abrasion and electrolytic action.

[B] 513.13.1 Materials. Control-air tubing shall be hard-drawn copper, Type L, ACR in accordance with ASTM B 42, ASTM B 43, ASTM B 68, ASTM B 88, ASTM B 251 and ASTM B 280. Fittings shall be wrought copper or brass, solder type in accordance with ASME B16.18 or ASME B16.22. Changes in direction shall be made with appropriate tool bends. Brass compression-type fittings shall be used at final connection to devices; other joints shall be brazed using a BCuP5 brazing alloy with solidus above 1,100°F (593°C) and liquids below 1,500°F (816°C). Brazing flux shall be used on copper-to-brass joints only.

Exception: Nonmetallic tubing used within control panels and at the final connection to devices provided all of the following conditions are met:

1. Tubing shall be listed by an approved agency for flame and smoke characteristics.

2. Tubing and connected device shall be completely enclosed within a galvanized or paint-grade steel enclosure of not less than 0.030 inch (0.76 mm) (No. 22 galvanized sheet gage) thickness. Entry to the enclosure shall be by copper tubing with a protective grommet of neoprene or teflon or by suitable brass compression to male barbed adapter.

3. Tubing shall be identified by appropriately documented coding.

4. Tubing shall be neatly tied and supported within the enclosure. Tubing bridging cabinets and doors or moveable devices shall be of sufficient length to avoid tension and excessive stress. Tubing shall be protected against abrasion. Tubing serving devices on doors shall be fastened along hinges.

[B] 513.13.2 Isolation from other functions. Control tubing serving other than smoke control functions shall be isolated by automatic isolation valves or shall be an independent system.

[B] 513.13.3 Testing. Test control-air tubing at three times the operating pressure for not less than 30 minutes without any noticeable loss in gauge pressure prior to final connection to devices.

[B] 513.14 Marking and identification. The detection and control systems shall be clearly marked at all junctions, accesses and terminations.

[F] 513.15 Control diagrams. Identical control diagrams shall be provided and maintained as required by the *International Fire Code.*

[F] 513.16 Fire fighter's smoke control panel. A fire fighter's smoke control panel for fire department emergency response

purposes only shall be provided in accordance with the *International Fire Code.*

[F] 513.17 System response time. Smoke control system activation shall comply with the *International Fire Code.*

[F] 513.18 Acceptance testing. Devices, equipment, components and sequences shall be tested in accordance with the *International Fire Code.*

[F] 513.19 System acceptance. Acceptance of the smoke control system shall be in accordance with the *International Fire Code.*

[B] 513.20 Underground building smoke exhaust system. Where required by the *International Building Code* for underground buildings, a smoke exhaust system shall be provided in accordance with this section.

[B] 513.20.1 Exhaust capability. Where compartmentation is required, each compartment shall have an independent, automatically activated smoke exhaust system capable of manual operation. The system shall have an air supply and smoke exhaust capability that will provide a minimum of six air changes per hour.

[F] 513.20.2 Operation. The smoke exhaust system shall be operated in accordance with the *International Fire Code.*

[F] 513.20.3 Alarm required. Activation of the smoke exhaust system shall activate an audible alarm at a constantly attended location in accordance with the *International Fire Code.*

SECTION 514
ENERGY RECOVERY VENTILATION SYSTEMS

514.1 General. Energy recovery ventilation systems shall be installed in accordance with this section. Where required for purposes of energy conservation, energy recovery ventilation systems shall also comply with the *International Energy Conservation Code.*

514.2 Prohibited applications. Energy recovery ventilation systems shall not be used in the following systems.

1. Hazardous exhaust systems covered in Section 510.

2. Dust, stock and refuse systems that convey explosive or flammable vapors, fumes or dust.

3. Smoke control systems covered in Section 513.

4. Commercial kitchen exhaust systems serving Type I and Type II hoods.

5. Clothes dryer exhaust systems covered in Section 504.

514.3 Access. A means of access shall be provided to the heat exchanger and other components of the system as required for service, maintenance, repair or replacement.

CHAPTER 6

DUCT SYSTEMS

SECTION 601
GENERAL

601.1 Scope. Duct systems used for the movement of air in air-conditioning, heating, ventilating and exhaust systems shall conform to the provisions of this chapter except as otherwise specified in Chapters 5 and 7.

Exception: Ducts discharging combustible material directly into any combustion chamber shall conform to the requirements of NFPA 82.

[B] 601.2 Air movement in egress elements. Exit access corridors shall not serve as supply, return, exhaust, relief or ventilation air ducts.

Exceptions:

1. Use of a corridor as a source of makeup air for exhaust systems in rooms that open directly onto such corridors, including toilet rooms, bathrooms, dressing rooms, smoking lounges and janitor closets, shall be permitted provided that each such corridor is directly supplied with outdoor air at a rate greater than the rate of makeup air taken from the corridor.

2. Where located within a dwelling unit, the use of corridors for conveying return air shall not be prohibited.

3. Where located within tenant spaces of 1,000 square feet (93 m²) or less in area, utilization of corridors for conveying return air is permitted.

[B] 601.2.1 Corridor ceiling. Use of the space between the corridor ceiling and the floor or roof structure above as a return air plenum is permitted for one or more of the following conditions:

1. The corridor is not required to be of fire-resistance-rated construction;

2. The corridor is separated from the plenum by fire-resistance-rated construction;

3. The air-handling system serving the corridor is shut down upon activation of the air-handling unit smoke detectors required by this code;

4. The air-handling system serving the corridor is shut down upon detection of sprinkler waterflow where the building is equipped throughout with an automatic sprinkler system; or

5. The space between the corridor ceiling and the floor or roof structure above the corridor is used as a component of an approved engineered smoke control system.

601.3 Contamination prevention. Exhaust ducts under positive pressure, chimneys, and vents shall not extend into or pass through ducts or plenums.

SECTION 602
PLENUMS

602.1 General. Supply, return, exhaust, relief and ventilation air plenums shall be limited to uninhabited crawl spaces, areas above a ceiling or below the floor, attic spaces and mechanical equipment rooms. Plenums shall be limited to one fire area. Fuel-fired appliances shall not be installed within a plenum.

602.2 Construction. Plenum enclosures shall be constructed of materials permitted for the type of construction classification of the building.

The use of gypsum boards to form plenums shall be limited to systems where the air temperatures do not exceed 125°F (52°C) and the building and mechanical system design conditions are such that the gypsum board surface temperature will be maintained above the airstream dew-point temperature. Air plenums formed by gypsum boards shall not be incorporated in air-handling systems utilizing evaporative coolers.

602.2.1 Materials exposed within plenums. Except as required by Sections 602.2.1.1 through 602.2.1.5, materials exposed within plenums shall be noncombustible or shall have a flame spread index of not more than 25 and a smoke-developed index of not more than 50 when tested in accordance with ASTM E 84.

Exceptions:

1. Rigid and flexible ducts and connectors shall conform to Section 603.

2. Duct coverings, linings, tape and connectors shall conform to Sections 603 and 604.

3. This section shall not apply to materials exposed within plenums in one- and two-family dwellings.

4. This section shall not apply to smoke detectors.

5. Combustible materials enclosed in approved gypsum board assemblies or enclosed in materials listed and labeled for such application.

602.2.1.1 Wiring. Combustible electrical or electronic wiring methods and materials, optical fiber cable, and optical fiber raceway exposed within a plenum shall have a peak optical density not greater than 0.50, an average optical density not greater than 0.15, and a flame spread not greater than 5 feet (1524 mm) when tested in accordance with NFPA 262. Only type OFNP (plenum rated nonconductive optical fiber cable) shall be installed in plenum-rated optical fiber raceways. Wiring, cable, and raceways addressed in this section shall be listed and labeled as plenum rated and shall be installed in accordance with *ICC Electrical Code.*

602.2.1.2 Fire sprinkler piping. Plastic fire sprinkler piping exposed within a plenum shall be used only in wet pipe systems and shall have a peak optical density not greater than 0.50, an average optical density not greater than 0.15, and a flame spread of not greater than 5 feet (1524 mm) when tested in accordance with UL 1887. Piping shall be listed and labeled.

602.2.1.3 Pneumatic tubing. Combustible pneumatic tubing exposed within a plenum shall have a peak optical density not greater than 0.50, an average optical density not greater than 0.15, and a flame spread of not greater than 5 feet (1524 mm) when tested in accordance with UL 1820. Combustible pneumatic tubing shall be listed and labeled.

602.2.1.4 Combustible electrical equipment. Combustible electrical equipment exposed within a plenum shall have a peak rate of heat release not greater than 100 kilowatts, a peak optical density not greater than 0.50 and an average optical density not greater than 0.15 when tested in accordance with UL 2043. Combustible electrical equipment shall be listed and labeled.

602.2.1.5 Foam plastic insulation. Foam plastic insulation used as wall or ceiling finish in plenums shall exhibit a flame spread index of 75 or less and a smoke developed index of 450 or less when tested in accordance with ASTM E 84 and shall also comply with Section 602.2.1.5.1, 602.2.1.5.2 or 602.2.1.5.3.

602.2.1.5.1 Separation required. The foam plastic insulation shall be separated from the plenum by a thermal barrier complying with Section 2603.4 of the *International Building Code*.

602.2.1.5.2 Approval. The foam plastic insulation shall be approved based on tests conducted in accordance with Section 2603.8 of the *International Building Code*.

602.2.1.5.3 Covering. The foam plastic insulation shall be covered by corrosion-resistant steel having a base metal thickness of not less than 0.0160 inch (0.4 mm).

602.3 Stud cavity and joist space plenums. Stud wall cavities and the spaces between solid floor joists to be utilized as air plenums shall comply with the following conditions:

1. Such cavities or spaces shall not be utilized as a plenum for supply air.

2. Such cavities or spaces shall not be part of a required fire-resistance-rated assembly.

3. Stud wall cavities shall not convey air from more than one floor level.

4. Stud wall cavities and joist space plenums shall comply with the floor penetration protection requirements of the *International Building Code*.

5. Stud wall cavities and joist space plenums shall be isolated from adjacent concealed spaces by approved fireblocking as required in the *International Building Code*.

[B] 602.4 Flood hazard. For structures located in flood hazard areas, plenum spaces shall be located above the design flood elevation or shall be designed and constructed to prevent water from entering or accumulating within the plenum spaces during floods up to the design flood elevation. If the plenum spaces are located below the design flood elevation, they shall be capable of resisting hydrostatic and hydrodynamic loads and stresses, including the effects of buoyancy, during the occurrence of flooding to the design flood elevation.

SECTION 603
DUCT CONSTRUCTION AND INSTALLATION

603.1 General. An air distribution system shall be designed and installed to supply the required distribution of air. The installation of an air distribution system shall not affect the fire protection requirements specified in the *International Building Code*. Ducts shall be constructed, braced, reinforced and installed to provide structural strength and durability.

603.2 Duct sizing. Ducts installed within a single dwelling unit shall be sized in accordance with ACCA Manual D or other approved methods. Ducts installed within all other buildings shall be sized in accordance with the ASHRAE *Handbook of Fundamentals* or other equivalent computation procedure.

603.3 Duct classification. Ducts shall be classified based on the maximum operating pressure of the duct at pressures of positive or negative 0.5, 1.0, 2.0, 3.0, 4.0, 6.0 or 10.0 inches of water column. The pressure classification of ducts shall equal or exceed the design pressure of the air distribution in which the ducts are utilized.

603.4 Metallic ducts. All metallic ducts shall be constructed as specified in the SMACNA *HVAC Duct Construction Standards—Metal and Flexible*.

Exception: Ducts installed within single dwelling units shall have a minimum thickness as specified in Table 603.4.

TABLE 603.4
DUCT CONSTRUCTION MINIMUM SHEET METAL THICKNESSES FOR SINGLE DWELLING UNITS

| DUCT SIZE | GALVANIZED | | APPROXIMATE ALUMINUM B&S GAGE |
	Minimum thickness (inches)	Equivalent Galvanized Gage No.	
Round ducts and enclosed Rectangular ducts			
14" or less	0.013	30	26
Over 14"	0.016	28	24
Exposed rectangular ducts			
14" or less	0.016	28	24
Over 14"	0.019	26	22

For SI:1 inch = 25.4 mm.

603.5 Nonmetallic ducts. Nonmetallic ducts shall be constructed with Class 0 or Class 1 duct material in accordance with UL 181. Fibrous duct construction shall conform to the SMACNA *Fibrous Glass Duct Construction Standards* or NAIMA *Fibrous Glass Duct Construction Standards*. The maximum air temperature within nonmetallic ducts shall not exceed 250°F (121°C).

603.5.1 Gypsum ducts. The use of gypsum boards to form air shafts (ducts) shall be limited to return air systems where the air temperatures do not exceed 125°F (52°C) and the gypsum board surface temperature is maintained above the airstream dew-point temperature. Air ducts formed by gypsum boards shall not be incorporated in air-handling systems utilizing evaporative coolers.

603.6 Flexible air ducts and flexible air connectors. Flexible air ducts, both metallic and nonmetallic, shall comply with Sections 603.6.1, 603.6.1.1, 603.6.3 and 603.6.4. Flexible air connectors, both metallic and nonmetallic, shall comply with Sections 606.6.2 through 603.6.4.

603.6.1 Flexible air ducts. Flexible air ducts, both metallic and nonmetallic, shall be tested in accordance with UL 181. Such ducts shall be listed and labeled as Class 0 or Class 1 flexible air ducts and shall be installed in accordance with Section 304.1.

603.6.1.1 Duct length. Flexible air ducts shall not be limited in length.

603.6.2 Flexible air connectors. Flexible air connectors, both metallic and nonmetallic, shall be tested in accordance with UL 181. Such connectors shall be listed and labeled as Class 0 or Class 1 flexible air connectors and shall be installed in accordance with Section 304.1.

603.6.2.1 Connector length. Flexible air connectors shall be limited in length to 14 feet (4267 mm).

603.6.2.2 Connector penetration limitations. Flexible air connectors shall not pass through any wall, floor or ceiling.

603.6.3 Air temperature. The design temperature of air to be conveyed in flexible air ducts and flexible air connectors shall be less than 250°F (121°C).

603.6.4 Flexible air duct and air connector clearance. Flexible air ducts and air connectors shall be installed with a minimum clearance to an appliance as specified in the appliance manufacturer's installation instructions.

603.7 Rigid duct penetrations. Duct system penetrations of walls, floors, ceilings and roofs and air transfer openings in such building components shall be protected as required by Section 607.

603.8 Underground ducts. Ducts shall be approved for underground installation. Metallic ducts not having an approved protective coating shall be completely encased in a minimum of 2 inches (51 mm) of concrete.

603.8.1 Slope. Ducts shall slope to allow drainage to a point provided with access.

603.8.2 Sealing. Ducts shall be sealed and secured prior to pouring the concrete encasement.

603.8.3 Plastic ducts and fittings. Plastic ducts shall be constructed of PVC having a minimum pipe stiffness of 8 psi (55 kPa) at 5-percent deflection when tested in accordance with ASTM D 2412. Plastic duct fittings shall be constructed of either PVC or high-density polyethylene. Plastic duct and fittings shall be utilized in underground installations only. The maximum design temperature for systems utilizing plastic duct and fittings shall be 150°F (66°C).

603.9 Joints, seams and connections. All longitudinal and transverse joints, seams and connections in metallic and non-

metallic ducts shall be constructed as specified in SMACNA HVAC *Duct Construction Standards—Metal and Flexible* and SMACNA *Fibrous Glass Duct Construction Standards* or NAIMA *Fibrous Glass Duct Construction Standards.* All longitudinal and transverse joints, seams and connections shall be sealed in accordance with the *International Energy Conservation Code.*

603.10 Supports. Ducts shall be supported with approved hangers at intervals not exceeding 10 feet (3048 mm) or by other approved duct support systems designed in accordance with the *International Building Code.* Flexible and other factory-made ducts shall be supported in accordance with the manufacturer's installation instructions.

603.11 Furnace connections. Ducts connecting to a furnace shall have a clearance to combustibles in accordance with the furnace manufacturer's installation instructions.

603.12 Condensation. Provisions shall be made to prevent the formation of condensation on the exterior of any duct.

[B] 603.13 Flood hazard areas. For structures in flood hazard areas, ducts shall be located above the design flood elevation or shall be designed and constructed to prevent water from entering or accumulating within the ducts during floods up to the design flood elevation. If the ducts are located below the design flood elevation, the ducts shall be capable of resisting hydrostatic and hydrodynamic loads and stresses, including the effects of buoyancy, during the occurrence of flooding to the design flood elevation.

603.14 Location. Ducts shall not be installed in or within 4 inches (102 mm) of the earth, except where such ducts comply with Section 603.7.

603.15 Mechanical protection. Ducts installed in locations where they are exposed to mechanical damage by vehicles or from other causes shall be protected by approved barriers.

603.16 Weather protection. All ducts including linings, coverings and vibration isolation connectors installed on the exterior of the building shall be adequately protected against the elements.

603.17 Registers, grilles and diffusers. Duct registers, grilles and diffusers shall be installed in accordance with the manufacturer's installation instructions. Balancing dampers or other means of supply air adjustment shall be provided in the branch ducts or at each individual duct register, grille or diffuser.

603.17.1 Floor registers. Floor registers shall resist, without structural failure, a 200-pound (90.8 kg) concentrated load on a 2-inch-diameter (51 mm) disc applied to the most critical area of the exposed face.

SECTION 604
INSULATION

604.1 General. Duct insulation shall conform to the requirements of Sections 604.2 through 604.13 and the *International Energy Conservation Code.*

604.2 Surface temperature. Ducts that operate at temperatures exceeding 120°F (49°C) shall have sufficient thermal insulation to limit the exposed surface temperature to 120°F (49°C).

604.3 Coverings and linings. Coverings and linings, including adhesives when used, shall have a flame spread index not more than 25 and a smoke-developed index not more than 50, when tested in accordance with ASTM E 84. Duct coverings and linings shall not flame, glow, smolder or smoke when tested in accordance with ASTM C 411 at the temperature to which they are exposed in service. The test temperature shall not fall below 250°F (121°C).

604.4 Foam plastic insulation. Foam plastic used as duct coverings and linings shall conform to the requirements of Section 604.

604.5 Appliance insulation. Listed and labeled appliances that are internally insulated shall be considered as conforming to the requirements of Section 604.

604.6 Penetration of assemblies. Duct coverings shall not penetrate a wall or floor required to have a fire-resistance rating or required to be fireblocked.

[EC] 604.7 Identification. External duct insulation and factory-insulated flexible duct shall be legibly printed or identified at intervals not greater than 36 inches (914 mm) with the name of the manufacturer, the thermal resistance R-value at the specified installed thickness and the flame spread and smoke-developed indexes of the composite materials. All duct insulation product R-values shall be based on insulation only, excluding air films, vapor retarders or other duct components, and shall be based on tested C-values at 75°F (24°C) mean temperature at the installed thickness, in accordance with recognized industry procedures. The installed thickness of duct insulation used to determine its R-values shall be determined as follows:

1. For duct board, duct liner and factory-made rigid ducts not normally subjected to compression, the nominal insulation thickness shall be used.

2. For duct wrap, the installed thickness shall be assumed to be 75 percent (25-percent compression) of nominal thickness.

3. For factory-made flexible air ducts, the installed thickness shall be determined by dividing the difference between the actual outside diameter and nominal inside diameter by two.

604.8 Lining installation. Linings shall be interrupted at the area of operation of a fire damper and at a minimum of 6 inches (152 mm) upstream of and 6 inches (152 mm) downstream of electric-resistance and fuel-burning heaters in a duct system. Metal nosings or sleeves shall be installed over exposed duct liner edges that face opposite the direction of airflow.

604.9 Thermal continuity. Where a duct liner has been interrupted, a duct covering of equal thermal performance shall be installed.

604.10 Service openings. Service openings shall not be concealed by duct coverings unless the exact location of the opening is properly identified.

604.11 Vapor retarders. Where ducts used for cooling are externally insulated, the insulation shall be covered with a vapor retarder having a maximum permeance of 0.05 perm [2.87 ng/(Pa · s · m²)] or aluminum foil having a minimum thickness

of 2 mils (0.051 mm). Insulations having a permeance of 0.05 perm [2.87 ng/(P · s · m²)] or less shall not be required to be covered. All joints and seams shall be sealed to maintain the continuity of the vapor retarder.

604.12 Weatherproof barriers. Insulated exterior ducts shall be protected with an approved weatherproof barrier.

604.13 Internal insulation. Materials used as internal insulation and exposed to the airstream in ducts shall be shown to be durable when tested in accordance with UL 181. Exposed internal insulation that is not impermeable to water shall not be used to line ducts or plenums from the exit of a cooling coil to the downstream end of the drain pan.

SECTION 605
AIR FILTERS

605.1 General. Heating and air-conditioning systems of the central type shall be provided with approved air filters. Filters shall be installed in the return air system, upstream from any heat exchanger or coil, in an approved convenient location. Liquid adhesive coatings used on filters shall have a flash point not lower than 325°F (163°C).

605.2 Approval. Media-type and electrostatic-type air filters shall be listed and labeled. Media-type air filters shall comply with UL 900. High efficiency particulate air filters shall comply with UL 586. Electrostatic-type air filters shall comply with UL 867. Air filters utilized within dwelling units shall be designed for the intended application and shall not be required to be listed and labeled.

605.3 Airflow over the filter. Ducts shall be constructed to allow an even distribution of air over the entire filter.

SECTION 606
SMOKE DETECTION SYSTEMS CONTROL

606.1 Controls required. Air distribution systems shall be equipped with smoke detectors listed and labeled for installation in air distribution systems, as required by this section.

606.2 Where required. Smoke detectors shall be installed where indicated in Sections 606.2.1 through 606.2.3.

> **Exception:** Smoke detectors shall not be required where air distribution systems are incapable of spreading smoke beyond the enclosing walls, floors and ceilings of the room or space in which the smoke is generated.

606.2.1 Return air systems. Smoke detectors shall be installed in return air systems with a design capacity greater than 2,000 cfm (0.9 m³/s), in the return air duct or plenum upstream of any filters, exhaust air connections, outdoor air connections, or decontamination equipment and appliances.

> **Exception:** Smoke detectors are not required in the return air system where all portions of the building served by the air distribution system are protected by area smoke detectors connected to a fire alarm system in accordance with the *International Fire Code*. The area smoke detection system shall comply with Section 606.4.

606.2.2 Common supply and return air systems. Where multiple air-handling systems share common supply or return air ducts or plenums with a combined design capacity greater than 2,000 cfm (0.9 m³/s), the return air system shall be provided with smoke detectors in accordance with Section 606.2.1.

Exception: Individual smoke detectors shall not be required for each fan-powered terminal unit, provided that such units do not have an individual design capacity greater than 2,000 cfm (0.9 m³/s) and will be shut down by activation of one of the following:

1. Smoke detectors required by Sections 606.2.1 and 606.2.3.

2. An approved area smoke detector system located in the return air plenum serving such units.

3. An area smoke detector system as prescribed in the exception to Section 606.2.1.

In all cases, the smoke detectors shall comply with Sections 606.4 and 606.4.1.

606.2.3 Return air risers. Where return air risers serve two or more stories and serve any portion of a return air system having a design capacity greater than 15,000 cfm (7.1 m³/s), smoke detectors shall be installed at each story. Such smoke detectors shall be located upstream of the connection between the return air riser and any air ducts or plenums.

606.3 Installation. Smoke detectors required by this section shall be installed in accordance with NFPA 72. The required smoke detectors shall be installed to monitor the entire airflow conveyed by the system including return air and exhaust or relief air. Access shall be provided to smoke detectors for inspection and maintenance.

606.4 Controls operation. Upon activation, the smoke detectors shall shut down the air distribution system. Air distribution systems that are part of a smoke control system shall switch to the smoke control mode upon activation of a detector.

606.4.1 Supervision. The duct smoke detectors shall be connected to a fire alarm system. The actuation of a duct smoke detector shall activate a visible and audible supervisory signal at a constantly attended location.

Exceptions:

1. The supervisory signal at a constantly attended location is not required where the duct smoke detector activates the building's alarm-indicating appliances.

2. In occupancies not required to be equipped with a fire alarm system, actuation of a smoke detector shall activate a visible and an audible signal in an approved location. Duct smoke detector trouble conditions shall activate a visible or audible signal in an approved location and shall be identified as air duct detector trouble.

[B] SECTION 607
DUCTS AND AIR TRANSFER OPENINGS

607.1 General. The provisions of this section shall govern the protection of duct penetrations and air transfer openings in fire-resistance-rated assemblies.

607.1.1 Ducts and air transfer openings without dampers. Ducts and air transfer openings that penetrate fire-resistance-rated assemblies and are not required to have dampers by this section shall comply with the requirements of Section 712 of the *International Building Code.*

607.2 Installation. Fire dampers, smoke dampers, combination fire/smoke dampers and ceiling dampers located within air distribution and smoke control systems shall be installed in accordance with the requirements of this section, and the manufacturer's installation instructions and listing.

607.2.1 Smoke control system. Where the installation of a fire damper will interfere with the operation of a required smoke control system in accordance with Section 513, approved alternative protection shall be utilized.

607.2.2 Hazardous exhaust ducts. Fire dampers for hazardous exhaust duct systems shall comply with Section 510.

607.3 Damper testing and ratings. Dampers shall be listed and bear the label of an approved testing agency indicating compliance with the standards in this section. Fire dampers shall comply with the requirements of UL 555. Only fire dampers labeled for use in dynamic systems shall be installed in heating, ventilation and air-conditioning systems designed to operate with fans on during a fire. Smoke dampers shall comply with the requirements of UL 555S. Combination fire/smoke dampers shall comply with the requirements of both UL 555 and UL 555S. Ceiling radiation dampers shall comply with the requirements of UL 555C.

607.3.1 Fire protection rating. Fire dampers shall have the minimum fire protection rating specified in Table 607.3.1 for the type of penetration.

607.3.1.1 Fire damper actuating device. The fire damper actuating device shall meet one of the following requirements:

1. The operating temperature shall be approximately 50°F (27.8°C) above the normal temperature within the duct system, but not less than 160°F (71°C).

2. The operating temperature shall be not more than 286°F (141°C) where located in a smoke control system complying with Section 513.

3. Where a combination fire/smoke damper is located in a smoke control system complying with Section 513, the operating temperature rating shall be approximately 50°F (27.8°C) above the maximum smoke control system designed operating temperature, or a maximum temperature of 350°F (177°C). The temperature shall not exceed the UL 555S degradation test temperature rating for a combination fire/smoke damper.

607.3.2 Smoke damper ratings. Smoke damper leakage ratings shall not be less than Class II. Elevated temperature ratings shall be not less than 250°F (121°C).

607.3.2.1 Smoke damper actuation methods. The smoke damper shall close upon actuation of a listed smoke detector or detectors installed in accordance with Section 607 of this code and Sections 907.10 and 907.11 of the *In-*

ternational Building Code and one of the following methods, as applicable:

1. Where a damper is installed within a duct, a smoke detector shall be installed in the duct within 5 feet (1524 mm) of the damper with no air outlets or inlets between the detector and the damper. The detector shall be listed for the air velocity, temperature and humidity anticipated at the point where it is installed. Other than in mechanical smoke control systems, dampers shall be closed upon fan shutdown where local smoke detectors require a minimum velocity to operate.

2. Where a damper is installed above smoke barrier doors in a smoke barrier, a spot-type detector listed for releasing service shall be installed on either side of the smoke barrier door opening.

3. Where a damper is installed within an unducted opening in a wall, a spot-type detector listed for releasing service shall be installed within 5 feet (1524 mm) horizontally of the damper.

4. Where a damper is installed in a corridor wall, the damper shall be permitted to be controlled by a smoke detection system installed in the corridor.

5. Where a total-coverage smoke detector system is provided within areas served by an HVAC system, dampers shall be permitted to be controlled by the smoke detection system.

TABLE 607.3.1
FIRE DAMPER RATING

TYPE OF PENETRATION	MINIMUM DAMPER RATING (hour)
Less than 3-hour fire-resistance-rated assemblies	$1^1/_2$
3-hour or greater fire-resistance-rated assembies	3

607.4 Access and identification. Fire and smoke dampers shall be provided with an approved means of access, large enough to permit inspection and maintenance of the damper and its operating parts. The access shall not affect the integrity of fire-resistance-rated assemblies. The access openings shall not reduce the fire-resistance rating of the assembly. Access points shall be permanently identified on the exterior by a label having letters not less than 0.5 inch (12.7 mm) in height reading: SMOKE DAMPER or FIRE DAMPER. Access doors in ducts shall be tight fitting and suitable for the required duct construction.

607.5 Where required. Fire dampers, smoke dampers, combination fire/smoke dampers and ceiling radiation dampers shall be provided at the locations prescribed in this section. Where an assembly is required to have both fire dampers and smoke dampers, combination fire/smoke dampers or a fire damper and a smoke damper shall be required.

607.5.1 Fire walls. Ducts and air transfer openings permitted in fire walls in accordance with Section 705.11 of the In-

ternational Building Code shall be protected with approved fire dampers installed in accordance with their listing.

607.5.2 Fire barriers. Duct penetrations and air transfer openings in fire barriers shall be protected with approved fire dampers installed in accordance with their listing.

Exceptions: Fire dampers are not required at penetrations of fire barriers where any of the following apply:

1. Penetrations are tested in accordance with ASTM E 119 as part of the fire-resistance-rated assembly.

2. Ducts are used as part of an approved smoke control system in accordance with Section 513.

3. Such walls are penetrated by ducted HVAC systems, have a required fire-resistance rating of 1 hour or less, are in areas of other than Group H and are in buildings equipped throughout with an automatic sprinkler system in accordance with Section 903.3.1.1 or 903.3.1.2 of the International Building Code. For the purposes of this exception, a ducted HVAC system shall be a duct system for the structure's HVAC system. Such a duct system shall be constructed of sheet metal not less than 26-gage (0.0217 inch) [0.55 mm] thickness and shall be continuous from the air-handling appliance or equipment to the air outlet and inlet terminals.

607.5.3 Fire partitions. Duct penetrations in fire partitions shall be protected with approved fire dampers installed in accordance with their listing.

Exceptions: In occupancies other than Group H, fire dampers are not required where any of the following apply:

1. The partitions are tenant separation and corridor walls in buildings equipped throughout with an automatic sprinkler system in accordance with Section 903.3.1.1 or 903.3.1.2 of the International Building Code and the duct is protected as a through penetration in accordance with the International Building Code.

2. The duct system is constructed of approved materials in accordance with this code and the duct penetrating the wall meets all of the following minimum requirements.

    2.1. The duct shall not exceed 100 square inches (0.06 m^2).

    2.2. The duct shall be constructed of steel a minimum of 0.0217-inch (0.55 mm) in thickness.

    2.3. The duct shall not have openings that communicate the corridor with adjacent spaces or rooms.

    2.4. The duct shall be installed above a ceiling.

    2.5. The duct shall not terminate at a wall register in the fire-resistance-rated wall.

    2.6. A minimum 12-inch-long (3048 mm) by 0.060-inch-thick (0.52 mm) steel sleeve shall be centered in each duct opening. The sleeve shall be secured to both sides of the wall and all four sides of the sleeve with minimum $1^1/_2$-inch by $1^1/_2$-inch by 0.060-inch (38 mm by 38 mm by 1.52 mm) steel retaining angles. The retaining angles shall be secured to the sleeve and the wall with No. 10

(M5) screws. The annular space between the steel sleeve and the wall opening shall be filled with rock (mineral) wool batting on all sides.

607.5.4 Corridors/smoke barriers. A listed smoke damper designed to resist the passage of smoke shall be provided at each point a duct or air transfer opening penetrates a smoke barrier wall or a corridor enclosure required to have smoke and draft control doors in accordance with the *International Building Code*. Smoke dampers and smoke damper actuation methods shall comply with Section 607.5.4.1.

Exceptions:

1. Smoke dampers are not required in corridor penetrations where the building is equipped throughout with an approved smoke control system in accordance with Section 513 and smoke dampers are not necessary for the operation and control of the system.

2. Smoke dampers are not required in smoke barrier penetrations where the openings in ducts are limited to a single smoke compartment and the ducts are constructed of steel.

3. Smoke dampers are not required in corridor penetrations where the duct is constructed of steel not less than 0.019 inch (0.48 mm) in thickness and there are no openings serving the corridor.

607.5.4.1 Smoke damper. The smoke damper shall close upon actuation of a listed smoke detector or detectors installed in accordance with the *International Building Code* and one of the following methods, as applicable:

1. Where a damper is installed within a duct, a smoke detector shall be installed in the duct within 5 feet (1524 mm) of the damper with no air outlets or inlets between the detector and the damper. The detector shall be listed for the air velocity, temperature and humidity anticipated at the point where it is installed.

2. Where a damper is installed above smoke barrier doors in a smoke barrier, a spot-type detector listed for releasing service shall be installed on either side of the smoke barrier door opening.

3. Where a damper is installed within an unducted opening in a wall, a spot-type detector listed for releasing service shall be installed within 5 feet (1524 mm) horizontally of the damper.

4. Where a damper is installed in a corridor wall, the damper shall be permitted to be controlled by a smoke detection system installed in the corridor.

5. Where a total-coverage smoke detector system is provided within all areas served by an HVAC system, dampers shall be permitted to be controlled by the smoke detection system.

607.5.5 Shaft enclosures. Ducts and air transfer openings shall not penetrate a shaft serving as an exit enclosure except as permitted by Section 1019.1.2 of the *International Building Code*.

607.5.5.1 Penetrations of shaft enclosures. Shaft enclosures that are permitted to be penetrated by ducts and air

transfer openings shall be protected with approved fire and smoke dampers installed in accordance with their listing.

Exceptions:

1. Fire dampers are not required at penetrations of shafts where:

    1.1. Steel exhaust subducts extend at least 22 inches (559 mm) vertically in exhaust shafts provided there is a continuous airflow upward to the outside, or

    1.2. Penetrations are tested in accordance with ASTM E 119 as part of the fire-resistance-rated assembly, or

    1.3. Ducts are used as part of an approved smoke control system designed and installed in accordance with Section 909 of the *International Building Code,* and where the fire damper will interfere with the operation of the smoke control system, or

    1.4. The penetrations are in parking garage exhaust or supply shafts that are separated from other building shafts by not less than 2-hour fire-resistance-rated construction.

2. In Group B occupancies, equipped throughout with an automatic sprinkler system in accordance with Section 903.3.1.1 of the *International Building Code*, smoke dampers are not required at penetrations of shafts where:

    2.1. Bathroom and toilet room exhaust openings with steel exhaust subducts, having a wall thickness of at least 0.019 inch (0.48 mm) extend at least 22 inches (559 mm) vertically and the exhaust fan at the upper terminus is powered continuously in accordance with the provisions of Section 909.11 of the *International Building Code*, and maintains airflow upward to the outside, or

    2.2. Ducts are used as part of an approved smoke control system designed and installed in accordance with Section 909 of the *International Building Code*, and where the smoke damper will interfere with the operation of the smoke control system.

3. Smoke dampers are not required at penetration of exhaust or supply shafts in parking garages that are separated from other building shafts by not less than 2-hour fire-resistance-rated construction.

607.6 Horizontal assemblies. Penetrations by air ducts of a floor, floor/ceiling assembly or the ceiling membrane of a roof/ceiling assembly shall be protected by a shaft enclosure that complies with the *International Building Code* or shall comply with this section.

607.6.1 Through penetrations. In occupancies other than Groups I-2 and I-3, a duct and air transfer opening system constructed of approved materials in accordance with this code that penetrates a fire-resistance-rated floor/ceiling assembly that connects not more than two stories is permitted without shaft enclosure protection provided a fire damper is installed at the floor line.

Exception: A duct is permitted to penetrate three floors or less without a fire damper at each floor provided it meets all of the following requirements.

1. The duct shall be contained and located within the cavity of a wall and shall be constructed of steel not less than 0.019 inch (0.48 mm) (26 gauge) in thickness.

2. The duct shall open into only one dwelling unit or sleeping unit and the duct system shall be continuous from the unit to the exterior of the building.

3. The duct shall not exceed 4-inch (102 mm) nominal diameter and the total area of such ducts shall not exceed 100 square inches (64 516 mm²) for any 100 square feet (9.3 m²) of the floor area.

4. The annular space around the duct is protected with materials that prevent the passage of flame and hot gases sufficient to ignite cotton waste where subjected to ASTM E 119 time-temperature conditions under a minimum positive pressure differential of 0.01 inch (2.49 Pa) of water at the location of the penetration for the time period equivalent to the fire-resistance rating of the construction penetrated.

5. Grille openings located in a ceiling of a fire-resistance-rated floor/ceiling or roof/ceiling assembly shall be protected with a ceiling radiation damper in accordance with Section 607.6.2.

607.6.2 Membrane penetrations. Where duct systems constructed of approved materials in accordance with this code penetrate a ceiling of a fire-resistance-rated floor/ceiling or roof/ceiling assembly, shaft enclosure protection is not required provided an approved ceiling radiation damper is installed at the ceiling line. Where a duct is not attached to a diffuser that penetrates a ceiling of a fire-resistance-rated floor/ceiling or roof/ceiling assembly, shaft enclosure protection is not required provided an approved ceiling radiation damper is installed at the ceiling line. Ceiling radiation dampers shall be installed in accordance with UL 555C and constructed in accordance with the details listed in a fire-resistance-rated assembly or shall be labeled to function as a heat barrier for air-handling outlet/inlet penetrations in the ceiling of a fire-resistance-rated assembly. Ceiling radiation dampers shall not be required where ASTM E 119 fire tests have shown that ceiling radiation dampers are not necessary in order to maintain the fire-resistance rating of the assembly.

607.6.3 Nonfire-resistance-rated assemblies. Duct systems constructed of approved materials in accordance with this code that penetrate nonfire-resistance-rated floor assemblies that connect not more than two stories are permitted without shaft enclosure protection provided that the annular space between the assembly and the penetrating duct is filled with an approved noncombustible material to resist the free passage of flame and the products of combustion. Duct systems constructed of approved materials in accordance with this code that penetrate non-rated floor assemblies that connect not more than three stories are permitted without shaft enclosure

protection provided that the annular space between the assembly and the penetrating duct is filled with an approved noncombustible material to resist the free passage of flame and the products of combustion, and a fire damper is installed at each floor line.

Exception: Fire dampers are not required in ducts within individual residential dwelling units.

607.7 Flexible ducts and air connectors. Flexible ducts and air connectors shall not pass through any fire-resistance-rated assembly.

CHAPTER 7
COMBUSTION AIR

SECTION 701
GENERAL

701.1 Scope. The provisions of this chapter shall govern the requirements for combustion and dilution air for fuel-burning appliances other than gas-fired appliances. The requirements for combustion and dilution air for gas-fired appliances shall be in accordance with the *International Fuel Gas Code*.

701.2 Combustion and dilution air required. Every room or space containing fuel-burning appliances shall be provided with combustion and dilution air as required by this code. Combustion and dilution air shall be provided in accordance with Section 702, 703, 704, 705, 706 or 707 or shall be provided by an approved engineered system. Direct vent appliances or equipment that do not draw combustion air from inside of the building are not required to be considered in the determination of the combustion and dilution air requirements. Combustion air requirements shall be determined based on the simultaneous operation of all fuel-burning appliances drawing combustion and dilution air from the room or space.

701.3 Circulation of air. The equipment and appliances within every room containing fuel-burning appliances shall be installed so as to allow free circulation of air. Provisions shall be made to allow for the simultaneous operation of mechanical exhaust systems, fireplaces or other equipment and appliances operating in the same room or space from which combustion and dilution air is being drawn. Such provisions shall prevent the operation of such appliances, equipment and systems from affecting the supply of combustion and dilution air.

701.4 Crawl space and attic space. For the purposes of this chapter, an opening to a naturally ventilated crawl space or attic space shall be considered equivalent to an opening to the outdoors.

701.4.1 Crawl space. Where lower combustion air openings connect with crawl spaces, such spaces shall have unobstructed openings to the outdoors at least twice that required for the combustion air openings. The height of the crawl space shall comply with the requirements of the *International Building Code* and shall be without obstruction to the free flow of air.

701.4.2 Attic space. Where combustion air is obtained from an attic area, the attic ventilating openings shall not be subject to ice or snow blockage, and the attic shall have not less than 30 inches (762 mm) vertical clear height at its maximum point. Attic ventilation openings shall be sufficient to provide the required volume of combustion air and the attic ventilation required by the *International Building Code*. The combustion air openings shall be provided with a sleeve of not less than 0.019-inch (0.5 mm) (No. 26 Gage) galvanized steel or other approved material extending from the appliance enclosure to at least 6 inches (152 mm) above the top of the ceiling joists and insulation.

701.5 Prohibited sources. Openings and ducts shall not connect appliance enclosures with a space in which the operation of a fan will adversely affect the flow of the combustion air. Combustion air shall not be obtained from a hazardous location, except where the fuel-fired appliances are located within the hazardous location and are installed in accordance with this code. Combustion air shall not be taken from a refrigeration machinery room, except where a refrigerant vapor detector system is installed to automatically shut off the combustion process in the event of refrigerant leakage. Combustion air shall not be obtained from any location below the design flood elevation.

SECTION 702
INSIDE AIR

702.1 All air from indoors. Combustion and dilution air shall be permitted to be obtained entirely from the indoors in buildings that are not of unusually tight construction. In buildings of unusually tight construction, combustion air shall be obtained from the outdoors in accordance with Section 703, 705, 706 or 707.

702.2 Air from the same room or space. The room or space containing fuel-burning appliances shall be an unconfined space as defined in Section 202.

702.3 Air from adjacent spaces. Where the volume of the room in which the fuel-burning appliances are located does not comply with Section 702.2, additional inside combustion and dilution air shall be obtained by opening the room to adjacent spaces so that the combined volume of all communicating spaces meets the volumetric requirement of Section 702.2. Openings connecting the spaces shall comply with Sections 702.3.1 and 702.3.2.

702.3.1 Number and location of openings. Two openings shall be provided, one within 1 foot (305 mm) of the ceiling of the room and one within 1 foot (305 mm) of the floor.

702.3.2 Size of openings. The net free area of each opening, calculated in accordance with Section 708, shall be a minimum of 1 square inch per 1,000 Btu/h (2201 mm²/kW) of input rating of the fuel-burning appliances drawing combustion and dilution air from the communicating spaces and shall be not less than 100 square inches (64 516 mm²).

SECTION 703
OUTDOOR AIR

703.1 All air from the outdoors. Where all combustion and dilution air is to be provided by outdoor air, the required combustion and dilution air shall be obtained by opening the room to the outdoors. Openings connecting the room to the outdoor air shall comply with Sections 703.1.1 through 703.1.4.

703.1.1 Number and location of openings. Two openings shall be provided, one within 1 foot (305 mm) of the ceiling of the room and one within 1 foot (305 mm) of the floor.

703.1.2 Size of direct openings. The net free area of each direct opening to the outdoors, calculated in accordance with Section 709, shall be a minimum of 1 square inch per 4,000 Btu/h (550 mm^2/kW) of combined input rating of the fuel-burning appliances drawing combustion and dilution air from the room.

703.1.3 Size of horizontal openings. The net free area of each opening, calculated in accordance with Section 709 and connected to the outdoors through a horizontal duct, shall be a minimum of 1 square inch per 2,000 Btu/h (1100 mm^2/kW) of combined input rating of the fuel-burning appliances drawing combustion and dilution air from the room. The cross-sectional area of the duct shall be equal to or greater than the required size of the opening.

703.1.4 Size of vertical openings. The net free area of each opening, calculated in accordance with Section 709 and connected to the outdoors through a vertical duct, shall be a minimum of 1 square inch per 4,000 Btu/h (550 mm^2/kW) of combined input rating of the fuel-burning appliances drawing combustion and dilution air from the room. The cross-sectional area of the duct shall be equal to or greater than the required size of the opening.

SECTION 704
COMBINED USE OF INSIDE AND OUTDOOR AIR (CONDITION 1)

704.1 Combination of air from inside and outdoors. This section shall apply only to appliances located in confined spaces in buildings not of unusually tight construction. Where the volumes of rooms and spaces are combined for the purpose of providing indoor combustion air, such rooms and spaces shall communicate through permanent openings in compliance with Sections 702.3.1 and 702.3.2. The required combustion and dilution air shall be obtained by opening the room to the outdoors using a combination of inside and outdoor air, prorated in accordance with Section 704.1.6. The ratio of interior spaces shall comply with Section 704.1.5. The number, location and ratios of openings connecting the space with the outdoor air shall comply with Sections 704.1.1 through 704.1.4.

704.1.1 Number and location of openings. At least two openings shall be provided, one within 1 foot (305 mm) of the ceiling of the room and one within 1 foot (305 mm) of the floor.

704.1.2 Ratio of direct openings. Where direct openings to the outdoors are provided in accordance with Section 703.1, the ratio of direct openings shall be the sum of the net free areas of both direct openings to the outdoors, divided by the sum of the required areas for both such openings as determined in accordance with Section 703.1.2.

704.1.3 Ratio of horizontal openings. Where openings connected to the outdoors through horizontal ducts are provided in accordance with Section 703.1, the ratio of horizontal openings shall be the sum of the net free areas of both

such openings, divided by the sum of the required areas for both such openings as determined in accordance with Section 703.1.3.

704.1.4 Ratio of vertical openings. Where openings connected to the outdoors through vertical ducts are provided in accordance with Section 703.1, the ratio of vertical openings shall be the sum of the net free areas of both such openings, divided by the sum of the required areas for both such openings as determined in accordance with Section 703.1.4.

704.1.5 Ratio of interior spaces. The ratio of interior spaces shall be the available volume of all communicating spaces, divided by the required volume as determined in accordance with Sections 702.2 and 702.3.

704.1.6 Prorating of inside and outdoor air. In spaces that utilize a combination of inside and outdoor air, the sum of the ratios of all direct openings, horizontal openings, vertical openings and interior spaces shall equal or exceed 1.

SECTION 705
COMBINED USE OF INSIDE AND OUTDOOR AIR (CONDITION 2)

705.1 General. This section shall apply only to appliances located in unconfined spaces in buildings of unusually tight construction. Combustion air supplied by a combined use of indoor and outdoor air shall be supplied through openings and ducts extending to the appliance room or to the vicinity of the appliance.

705.1.1 Openings and supply ducts. Openings shall be provided, located and sized in accordance with Sections 702.3.1 and 702.3.2; additionally, there shall be one opening to the outdoors having a free area of at least 1 square inch per 5,000 Btu/h (440 mm^2/kW) of total input of all appliances in the space.

SECTION 706
FORCED COMBUSTION AIR SUPPLY

706.1 General. Where all combustion air and dilution air is provided by a mechanical forced-air system, the combustion air and dilution air shall be supplied at the minimum rate of 1 cfm per 2,400 Btu/h [0.00067 m^3/(s · kW)] of combined input rating of all the fuel-burning appliances served. Each of the appliances served shall be electrically interlocked to the mechanical forced-air system so as to prevent operation of the appliances when the mechanical system is not in operation. Where combustion air and dilution air is provided by the building's mechanical ventilation system, the system shall provide the specified combustion/dilution air rate in addition to the required ventilation air.

SECTION 707
DIRECT CONNECTION

707.1 General. Fuel-burning appliances that are listed and labeled for direct combustion air connection to the outdoors shall be installed in accordance with the manufacturer's installation instructions.

SECTION 708
COMBUSTION AIR DUCTS

708.1 General. Combustion air ducts shall:

1. Be of galvanized steel complying with Chapter 6 or of equivalent corrosion-resistant material approved for this application.

   Exception: Within dwelling units, unobstructed stud and joist spaces shall not be prohibited from conveying combustion air, provided that not more than one required fireblock is removed.

2. Have a minimum cross-sectional dimension of 3 inches (76 mm).

3. Terminate in an unobstructed space allowing free movement of combustion air to the appliances.

4. Have the same cross-sectional areas as the free area of the openings to which they connect.

5. Serve a single appliance enclosure.

6. Not serve both upper and lower combustion air openings where both such openings are used. The separation between ducts serving upper and lower combustion air openings shall be maintained to the source of combustion air.

7. Not be screened where terminating in an attic space.

8. Not slope downward toward the source of combustion air, where serving the upper required combustion air opening.

SECTION 709
OPENING OBSTRUCTIONS

709.1 General. The required size of openings for combustion and dilution air shall be based on the net free area of each opening. The net free area of an opening shall be that specified by the manufacturer of the opening covering. In the absence of such information, openings covered with metal louvers shall be deemed to have a net free area of 75 percent of the area of the opening, and openings covered with wood louvers shall be deemed to have a net free area of 25 percent of the area of the opening. Louvers and grills shall be fixed in the open position.

Exception: Louvers interlocked with the appliance so that they are proven to be in the full open position prior to main burner ignition and during main burner operation. Means shall be provided to prevent the main burner from igniting if the louvers fail to open during burner startup and to shut down the main burner if the louvers close during operation.

709.2 Dampered openings. Where the combustion air openings are provided with volume, smoke or fire dampers, the dampers shall be electrically interlocked with the firing cycle of the appliances served, so as to prevent operation of any appliance that draws combustion and dilution air from the room when any of the dampers are closed. Manually operated dampers shall not be installed in combustion air openings.

SECTION 710
OPENING LOCATION AND PROTECTION

710.1 General. Combustion air openings to the outdoors shall comply with the location and protection provisions of Sections 401.5 and 401.6 applicable to outside air intake openings.

CHAPTER 8
CHIMNEYS AND VENTS

SECTION 801
GENERAL

801.1 Scope. This chapter shall govern the installation, maintenance, repair and approval of factory-built chimneys, chimney liners, vents and connectors. This chapter shall also govern the utilization of masonry chimneys. Gas-fired appliances shall be vented in accordance with the *International Fuel Gas Code.*

801.2 General. Every fuel-burning appliance shall discharge the products of combustion to a vent, factory-built chimney or masonry chimney, except for appliances vented in accordance with Section 804. The chimney or vent shall be designed for the type of appliance being vented.

801.2.1 Oil-fired appliances. Oil-fired appliances shall be vented in accordance with this code and NFPA 31.

801.3 Masonry chimneys. Masonry chimneys shall be constructed in accordance with the *International Building Code.*

801.4 Positive flow. Venting systems shall be designed and constructed so as to develop a positive flow adequate to convey all combustion products to the outside atmosphere.

801.5 Design. Venting systems shall be designed in accordance with this chapter or shall be approved engineered systems.

801.6 Minimum size of chimney or vent. Except as otherwise provided for in this chapter, the size of the chimney or vent, serving a single appliance, except engineered systems, shall have a minimum area equal to the area of the appliance connection.

801.7 Solid fuel appliance flues. The cross-sectional area of a flue serving a solid fuel-burning appliance shall be not greater than three times the cross-sectional area of the appliance flue collar or flue outlet.

801.8 Abandoned inlet openings. Abandoned inlet openings in chimneys and vents shall be closed by an approved method.

801.9 Positive pressure. Where an appliance equipped with a forced or induced draft system creates a positive pressure in the venting system, the venting system shall be designed and listed for positive pressure applications.

801.10 Connection to fireplace. Connection of appliances to chimney flues serving fireplaces shall be in accordance with Sections 801.10.1 through 801.10.3.

801.10.1 Closure and access. A noncombustible seal shall be provided below the point of connection to prevent entry of room air into the flue. Means shall be provided for access to the flue for inspection and cleaning.

801.10.2 Connection to factory-built fireplace flue. An appliance shall not be connected to a flue serving a factory-built fireplace unless the appliance is specifically listed for such installation. The connection shall be made in accordance with the appliance manufacturer's installation instructions.

801.10.3 Connection to masonry fireplace flue. A connector shall extend from the appliance to the flue serving a masonry fireplace such that the flue gases are exhausted directly into the flue. The connector shall be provided with access or shall be removable for inspection and cleaning of both the connector and the flue. Listed direct connection devices shall be installed in accordance with their listing.

801.11 Multiple solid fuel prohibited. A solid fuel-burning appliance or fireplace shall not connect to a chimney passageway venting another appliance.

801.12 Chimney entrance. Connectors shall connect to a chimney flue at a point not less than 12 inches (305 mm) above the lowest portion of the interior of the chimney flue.

801.13 Cleanouts. Masonry chimney flues shall be provided with a cleanout opening having a minimum height of 6 inches (152 mm). The upper edge of the opening shall be located not less than 6 inches (152 mm) below the lowest chimney inlet opening. The cleanout shall be provided with a tight-fitting, noncombustible cover.

Exception: Cleanouts shall not be required for chimney flues serving masonry fireplaces, if such flues are provided with access through the fireplace opening.

801.14 Connections to exhauster. All appliance connections to a chimney or vent equipped with a power exhauster shall be made on the inlet side of the exhauster. All joints and piping on the positive pressure side of the exhauster shall be listed for positive pressure applications as specified by the manufacturer's installation instructions for the exhauster.

801.15 Fuel-fired appliances. Masonry chimneys utilized to vent fuel-fired appliances shall be located, constructed and sized as specified in the manufacturer's installation instructions for the appliances being vented.

801.16 Flue lining. Masonry chimneys shall be lined. The lining material shall be compatible with the type of appliance connected, in accordance with the appliance listing and manufacturer's installation instructions. Listed materials used as flue linings shall be installed in accordance with their listings and the manufacturer's installation instructions.

801.16.1 Residential and low-heat appliances (general). Flue lining systems for use with residential-type and low-heat appliances shall be limited to the following:

1. Clay flue lining complying with the requirements of ASTM C 315 or equivalent. Clay flue lining shall be installed in accordance with the *International Building Code.*

2. Listed chimney lining systems complying with UL 1777.

3. Other approved materials that will resist, without cracking, softening or corrosion, flue gases and condensate at temperatures up to 1,800°F (982°C).

801.17 Space around lining. The space surrounding a flue lining system or other vent installed within a masonry chimney shall not be used to vent any other appliance. This shall not prevent the installation of a separate flue lining in accordance with the manufacturer's installation instructions and this code.

801.18 Existing chimneys and vents. Where an appliance is permanently disconnected from an existing chimney or vent, or where an appliance is connected to an existing chimney or vent during the process of a new installation, the chimney or vent shall comply with Sections 801.18.1 through 801.18.4.

801.18.1 Size. The chimney or vent shall be resized as necessary to control flue gas condensation in the interior of the chimney or vent and to provide the appliance or appliances served with the required draft. For the venting of oil-fired appliances to masonry chimneys, the resizing shall be in accordance with NFPA 31.

801.18.2 Flue passageways. The flue gas passageway shall be free of obstructions and combustible deposits and shall be cleaned if previously used for venting a solid or liquid fuel-burning appliance or fireplace. The flue liner, chimney inner wall or vent inner wall shall be continuous and shall be free of cracks, gaps, perforations or other damage or deterioration which would allow the escape of combustion products, including gases, moisture and creosote. Where an oil-fired appliance is connected to an existing masonry chimney, such chimney flue shall be repaired or relined in accordance with NFPA 31.

801.18.3 Cleanout. Masonry chimneys shall be provided with a cleanout opening complying with Section 801.13.

801.18.4 Clearances. Chimneys and vents shall have air-space clearance to combustibles in accordance with the *International Building Code* and the chimney or vent manufacturer's installation instructions.

Exception: Masonry chimneys equipped with a chimney lining system tested and listed for installation in chimneys in contact with combustibles in accordance with UL 1777, and installed in accordance with the manufacturer's instructions, shall not be required to have clearance between combustible materials and exterior surfaces of the masonry chimney. Noncombustible fireblocking shall be provided in accordance with the *International Building Code*.

801.19 Multistory prohibited. Common venting systems for appliances located on more than one floor level shall be prohibited, except where all of the appliances served by the common vent are located in rooms or spaces that are accessed only from the outdoors. The appliance enclosures shall not communicate with the occupiable areas of the building.

801.20 Plastic vent joints. Plastic pipe and fittings used to vent appliances shall be installed in accordance with the pipe manufacturer's installation instructions and the appliance manufacturer's installation instructions. Solvent cement joints between ABS pipe and fittings shall be cleaned. Solvent cement joints between CPVC and PVC pipe and fittings shall be primed. The primer shall be a contrasting color.

Exception: Where compliance with this section would conflict with the appliance manufacturer's installation instructions.

SECTION 802
VENTS

802.1 General. All vent systems shall be listed and labeled. Type L vents and pellet vents shall be tested in accordance with UL 641.

802.2 Vent application. The application of vents shall be in accordance with Table 802.2.

TABLE 802.2
VENT APPLICATION

VENT TYPES	APPLIANCE TYPES
Type L oil vents	Oil-burning appliances listed and labeled for venting with Type L vents; gas appliances listed and labeled for venting with Type B vents.
Pellet vents	Pellet fuel-burning appliances listed and Labeled for venting with pellet vents.

802.3 Installation. Vent systems shall be sized, installed and terminated in accordance with the vent and appliance manufacturer's installation instructions.

802.4 Vent termination caps required. Type L vents shall terminate with a listed and labeled cap in accordance with the vent manufacturer's installation instructions.

802.5 Type L vent terminations. Type L vents shall terminate not less than 2 feet (610 mm) above the highest point of the roof penetration and not less than 2 feet (610 mm) higher than any portion of a building within 10 feet (3048 mm).

802.6 Minimum vent heights. Vents shall terminate not less than 5 feet (1524 mm) in vertical height above the highest connected appliance flue collar.

Exceptions:

1. Venting systems of direct vent appliances shall be installed in accordance with the appliance and the vent manufacturer's instructions.

2. Appliances listed for outdoor installations incorporating integral venting means shall be installed in accordance with their listings and the manufacturer's installation instructions.

3. Pellet vents shall be installed in accordance with the appliance and the vent manufacturer's installation instructions.

802.7 Support of vents. All portions of vents shall be adequately supported for the design and weight of the materials employed.

802.8 Insulation shield. Where vents pass through insulated assemblies, an insulation shield constructed of not less than No. 26 Gage sheet metal shall be installed to provide clearance between the vent and the insulation material. The clearance shall be not less than the clearance to combustibles specified by the vent manufacturer's installation instructions. Where vents pass through attic space, the shield shall terminate not less than 2 inches (51 mm) above the insulation materials and shall be secured in place to prevent displacement. Insulation shields provided as part of a listed vent system shall be installed in accordance with the manufacturer's installation instructions.

SECTION 803
CONNECTORS

803.1 Connectors required. Connectors shall be used to connect appliances to the vertical chimney or vent, except where the chimney or vent is attached directly to the appliance.

803.2 Location. Connectors shall be located entirely within the room in which the connecting appliance is located, except as provided for in Section 803.10.4. Where passing through an unheated space, a connector shall not be constructed of single-wall pipe.

803.3 Size. The connector shall not be smaller than the size of the flue collar supplied by the manufacturer of the appliance. Where the appliance has more than one flue outlet, and in the absence of the manufacturer's specific instructions, the connector area shall be not less than the combined area of the flue outlets for which it acts as a common connector.

803.4 Branch connections. All branch connections to the vent connector shall be made in accordance with the vent manufacturer's instructions.

803.5 Manual dampers. Manual dampers shall not be installed in connectors except in chimney connectors serving solid fuel-burning appliances.

803.6 Automatic dampers. Automatic dampers shall be listed and labeled in accordance with UL 17 for oil-fired heating appliances. The dampers shall be installed in accordance with the manufacturer's installation instructions. An automatic vent damper device shall not be installed on an existing appliance unless the appliance is listed and labeled and the device is installed in accordance with the terms of its listing. The name of the installer and date of installation shall be marked on a label affixed to the damper device.

803.7 Connectors serving two or more appliances. Where two or more connectors enter a common vent or chimney, the smaller connector shall enter at the highest level consistent with available headroom or clearance to combustible material.

803.8 Vent connector construction. Vent connectors shall be constructed of metal. The minimum nominal thickness of the connector shall be 0.019 inch (0.5 mm) (No. 28 Gage) for galvanized steel, 0.022 inch (0.6 mm) (No. 26 B & S Gage) for copper, and 0.020 inch (0.5 mm) (No. 24 B & S Gage) for aluminum.

803.9 Chimney connector construction. Chimney connectors for low-heat appliances shall be of sheet steel pipe having resistance to corrosion and heat not less than that of galvanized steel specified in Table 803.9(1). Connectors for medium-heat appliances and high-heat appliances shall be of sheet steel not less than the thickness specified in Table 803.9(2).

TABLE 803.9(1)
MINIMUM CHIMNEY CONNECTOR THICKNESS FOR
LOW-HEAT APPLIANCES

DIAMETER OF CONNECTOR (inches)	MINIMUM NOMINAL THICKNESS (galvanized) (inches)
5 and smaller	0.022 (No. 26 Gage)
Larger than 5 and up to 10	0.028 (No. 24 Gage)
Larger than 10 and up to 16	0.034 (No. 22 Gage)
Larger than 16	0.064 (No. 16 Gage)

For SI:1 inch = 25.4 mm.

TABLE 803.9(2)
MINIMUM CHIMNEY CONNECTOR THICKNESS FOR
MEDIUM- AND HIGH-HEAT APPLIANCES

AREA (square inches)	EQUIVALENT ROUND DIAMETER (inches)	MINIMUM NOMINAL THICKNESS (inches)
0-154	0-14	0.060 (No. 16 Gage)
155-201	15-16	0.075 (No. 14 Gage)
202-254	17-18	0.105 (No. 12 Gage)
Greater than 254	Greater than 18	0.135 (No. 10 Gage)

For SI:1 inch = 25.4 mm, 1 square inch = 645.16 mm^2.

803.10 Installation. Connectors shall be installed in accordance with Sections 803.10.1 through 803.10.6.

803.10.1 Supports and joints. Connectors shall be supported in an approved manner, and joints shall be fastened with sheet metal screws, rivets or other approved means.

803.10.2 Length. The maximum horizontal length of a single-wall connector shall be 75 percent of the height of the chimney or vent.

803.10.3 Connection. The connector shall extend to the inner face of the chimney or vent liner, but not beyond. A connector entering a masonry chimney shall be cemented to masonry in an approved manner. Where thimbles are installed to facilitate removal of the connector from the masonry chimney, the thimble shall be permanently cemented in place with high-temperature cement.

803.10.4 Connector pass-through. Chimney connectors shall not pass through any floor or ceiling, nor through a fire-resistance-rated wall assembly. Chimney connectors for domestic-type appliances shall not pass through walls or partitions constructed of combustible material to reach a masonry chimney unless:

1. The connector is labeled for wall pass-through and is installed in accordance with the manufacturer's instructions; or

2. The connector is put through a device labeled for wall pass-through; or

3. The connector has a diameter not larger than 10 inches (254 mm) and is installed in accordance with one of the methods in Table 803.10.4. Concealed metal parts of the pass-through system in contact with flue gases shall be of stainless steel or equivalent material that resists corrosion, softening or cracking up to 1,800°F (980°C).

TABLE 803.10.4
CHIMNEY CONNECTOR SYSTEMS AND CLEARANCES TO COMBUSTIBLE WALL MATERIALS FOR DOMESTIC HEATING APPLIANCES[a,b,c,d]

System A (12-inch clearance)	A 3.5-inch-thick brick wall shall be framed into the combustible wall. A 0.625-inch-thick fire-clay liner (ASTM C 315 or equivalent)[e] shall be firmly cemented in the center of the brick wall maintaining a 12-inch clearance to combustibles. The clay liner shall run from the outer surface of the bricks to the inner surface of the chimney liner.
System B (9-inch clearance)	A labeled solid-insulated factory-built chimney section (1-inch insulation) the same inside diameter as the connector shall be utilized. Sheet metal supports cut to maintain a 9-inch clearance to combustibles shall be fastened to the wall surface and to the chimney section. Fasteners shall not penetrate the chimney flue liner. The chimney length shall be flush with the masonry chimney liner and sealed to the masonry with water-insoluble refractory cement. Chimney manufacturers' parts shall be utilized to securely fasten the chimney connector to the chimney section.
System C (6-inch clearance)	A sheet metal (minimum number 24 Gage) ventilated thimble having two 1-inch air channels shall be installed with a sheet steel chimney connector (minimum number 24 Gage). Sheet steel supports (minimum number 24 Gage) shall be cut to maintain a 6-inch clearance between the thimble and combustibles. One side of the support shall be fastened to the wall on all sides. Glass-fiber insulation shall fill the 6-inch space between the thimble and the supports.
System D (2-inch clearance)	A labeled solid-insulated factory-built chimney section (1-inch insulation) with a diameter 2 inches larger than the chimney connector shall be installed with a sheet steel chimney connector (minimum number 24 Gage). Sheet metal supports shall be positioned to maintain a 2-inch clearance to combustibles and to hold the chimney connector to ensure that a 1-inch airspace surrounds the chimney connector through the chimney section. The steel support shall be fastened to the wall on all sides and the chimney section shall be fastened to the supports. Fasteners shall not penetrate the liner of the chimney section.

For SI: 1 inch = 25.4 mm, 1.0 Btu × in/ft² • h • °F = 0.144 W/m² • °K.

a. Insulation material that is part of the wall pass-through system shall be noncombustible and shall have a thermal conductivity of 1.0 Btu × in/ft² • h • °F or less.

b. All clearances and thicknesses are minimums.

c. Materials utilized to seal penetrations for the connector shall be noncombustible.

d. Connectors for all systems except System B shall extend through the wall pass-through system to the inner face of the flue liner.

e. ASTM C 315.

803.10.5 Pitch. Connectors shall rise vertically to the chimney or vent with a minimum pitch equal to one-fourth unit vertical in 12 units horizontal (2-percent slope).

803.10.6 Clearances. Connectors shall have a minimum clearance to combustibles in accordance with Table 803.10.6. The clearances specified in Table 803.10.6 apply, except where the listing and labeling of an appliance specifies a different clearance, in which case the labeled clearance shall apply. The clearance to combustibles for connectors shall be reduced only in accordance with Section 308.

TABLE 803.10.6
CONNECTOR CLEARANCES TO COMBUSTIBLES

TYPE OF APPLIANCE	MINIMUM CLEARANCE (inches)
Comestic-type appliances	
Chimney and vent connectors	
Electric and oil incinerators	18
Oil and solid fuel appliances	18
Oil appliances labeled for venting with Type L vents	9
Commercial, industrial-type appliances	
Low-heat appliances	
Chimney connectors	
Oil and solid fuel boilers, furnaces and water heaters	18
Oil unit heaters	18
Other low-heat industrial appliances	18
Medium-heat appliances	
Chimney connectors	
All oil and solid fuel appliances	36
High-heat appliances Masonry or metal connectors All oil and solid fuel appliances	(As determined by the code official)

For SI: 1 inch = 25.4 mm.

SECTION 804
DIRECT-VENT, INTEGRAL VENT AND MECHANICAL DRAFT SYSTEMS

804.1 Direct-vent terminations. Vent terminals for direct-vent appliances shall be installed in accordance with the manufacturer's installation instructions

804.2 Appliances with integral vents. Appliances incorporating integral venting means shall be installed in accordance with their listings and the manufacturer's installation instructions.

804.2.1 Terminal clearances. Appliances designed for natural draft venting and incorporating integral venting means shall be located so that a minimum clearance of 9 inches

(229 mm) is maintained between vent terminals and from any openings through which combustion products enter the building. Appliances using forced draft venting shall be located so that a minimum clearance of 12 inches (305 mm) is maintained between vent terminals and from any openings through which combustion products enter the building.

804.3 Mechanical draft systems. Mechanical draft systems of either forced or induced draft design shall comply with Sections 804.3.1 through 804.3.7.

804.3.1 Forced draft systems. Forced draft systems and all portions of induced draft systems under positive pressure during operation shall be designed and installed so as to be gas tight to prevent leakage of combustion products into a building.

804.3.2 Automatic shutoff. Power exhausters serving automatically fired appliances shall be electrically connected to each appliance to prevent operation of the appliance when the power exhauster is not in operation.

804.3.3 Termination. The termination of chimneys or vents equipped with power exhausters shall be located a minimum of 10 feet (3048 mm) from the lot line or from adjacent buildings. The exhaust shall be directed away from the building.

804.3.4 Horizontal terminations. Horizontal terminations shall comply with the following requirements:

1. Where located adjacent to walkways, the termination of mechanical draft systems shall be not less than 7 feet (2134 mm) above the level of the walkway.

2. Vents shall terminate at least 3 feet (914 mm) above any forced air inlet located within 10 feet (3048 mm).

3. The vent system shall terminate at least 4 feet (1219 mm) below, 4 feet (1219 mm) horizontally from or 1 foot (305 mm) above any door, window or gravity air inlet into the building.

4. The vent termination point shall not be located closer than 3 feet (914 mm) to an interior corner formed by two walls perpendicular to each other.

5. The vent termination shall not be mounted directly above or within 3 feet (914 mm) horizontally from an oil tank vent or gas meter.

6. The bottom of the vent termination shall be located at least 12 inches (305 mm) above finished grade.

804.3.5 Vertical terminations. Vertical terminations shall comply with the following requirements:

1. Where located adjacent to walkways, the termination of mechanical draft systems shall be not less than 7 feet (2134 mm) above the level of the walkway.

2. Vents shall terminate at least 3 feet (914 mm) above any forced air inlet located within 10 feet (3048 mm).

3. Where the vent termination is located below an adjacent roof structure, the termination point shall be located at least 3 feet (914 mm) from such structure.

4. The vent shall terminate at least 4 feet (1219 mm) below, 4 feet (1219 mm) horizontally from, or 1 foot (305 mm) above any door, window or gravity air inlet for the building.

5. A vent cap shall be installed to prevent rain from entering the vent system.

6. The vent termination shall be located at least 3 feet (914 mm) horizontally from any portion of the roof structure.

804.3.6 Exhauster connections. An appliance vented by natural draft shall not be connected into a vent, chimney or vent connector on the discharge side of a mechanical flue exhauster.

804.3.7 Exhauster sizing. Mechanical flue exhausters and the vent system served shall be sized and installed in accordance with the manufacturer's installation instructions.

804.3.8 Mechanical draft systems for manually fired appliances and fireplaces. A mechanical draft system shall be permitted to be used with manually fired appliances and fireplaces where such system complies with all of the following requirements:

1. The mechanical draft device shall be listed and installed in accordance with the manufacturer's installation instructions.

2. A device shall be installed that produces visible and audible warning upon failure of the mechanical draft device or loss of electrical power, at any time that the mechanical draft device is turned on. This device shall be equipped with a battery backup if it receives power from the building wiring.

3. A smoke detector shall be installed in the room with the appliance or fireplace. This device shall be equipped with a battery backup if it receives power from the building wiring.

SECTION 805
FACTORY-BUILT CHIMNEYS

805.1 Listing. Factory-built chimneys shall be listed and labeled and shall be installed and terminated in accordance with the manufacturer's installation instructions.

805.2 Solid fuel appliances. Factory-built chimneys for use with solid fuel-burning appliances shall comply with the Type HT requirements of UL 103.

> **Exception:** Chimneys for use with fireplace stoves listed only to UL 737 shall comply with the requirements of UL 103.

805.3 Factory-built fireplaces. Chimneys for use with factory-built fireplaces shall comply with the requirements of UL 127.

805.4 Support. Where factory-built chimneys are supported by structural members, such as joists and rafters, such members shall be designed to support the additional load.

805.5 Medium-heat appliances. Factory-built chimneys for medium-heat appliances producing flue gases having a temperature above 1,000°F (538°C), measured at the entrance to the chimney, shall comply with UL 959.

805.6 Decorative shrouds. Decorative shrouds shall not be installed at the termination of factory-built chimneys except where such shrouds are listed and labeled for use with the specific factory-built chimney system and are installed in accordance with Section 304.1.

SECTION 806
METAL CHIMNEYS

806.1 General. Metal chimneys shall be constructed and installed in accordance with NFPA 211.

CHAPTER 9

SPECIFIC APPLIANCES, FIREPLACES AND SOLID FUEL-BURNING EQUIPMENT

SECTION 901
GENERAL

901.1 Scope. This chapter shall govern the approval, design, installation, construction, maintenance, alteration and repair of the appliances and equipment specifically identified herein and factory-built fireplaces. The approval, design, installation, construction, maintenance, alteration and repair of gas-fired appliances shall be regulated by the *International Fuel Gas Code*.

901.2 General. The requirements of this chapter shall apply to the mechanical equipment and appliances regulated by this chapter, in addition to the other requirements of this code.

901.3 Hazardous locations. Fireplaces and solid fuel-burning appliances shall not be installed in hazardous locations.

901.4 Fireplace accessories. Listed fireplace accessories shall be installed in accordance with the conditions of the listing and the manufacturer's installation instructions.

SECTION 902
MASONRY FIREPLACES

902.1 General. Masonry fireplaces shall be constructed in accordance with the *International Building Code*.

SECTION 903
FACTORY-BUILT FIREPLACES

903.1 General. Factory-built fireplaces shall be listed and labeled and shall be installed in accordance with the conditions of the listing. Factory-built fireplaces shall be tested in accordance with UL 127.

903.2 Hearth extensions. Hearth extensions of approved factory-built fireplaces and fireplace stoves shall be installed in accordance with the listing of the fireplace. The hearth extension shall be readily distinguishable from the surrounding floor area.

903.3 Unvented gas log heaters. An unvented gas log heater shall not be installed in a factory-built fireplace unless the fireplace system has been specifically tested, listed and labeled for such use in accordance with UL 127.

SECTION 904
PELLET FUEL-BURNING APPLIANCES

904.1 General. Pellet fuel-burning appliances shall be listed and labeled and shall be installed in accordance with the terms of the listing.

SECTION 905
FIREPLACE STOVES AND ROOM HEATERS

905.1 General. Fireplace stoves and solid-fuel-type room heaters shall be listed and labeled and shall be installed in accordance with the conditions of the listing. Fireplace stoves shall be tested in accordance with UL 737. Solid-fuel-type room heaters shall be tested in accordance with UL 1482. Fireplace inserts intended for installation in fireplaces shall be listed and labeled in accordance with the requirements of UL 1482 and shall be installed in accordance with the manufacturer's installation instructions.

905.2 Connection to fireplace. The connection of solid fuel appliances to chimney flues serving fireplaces shall comply with Sections 801.7 and 801.10.

SECTION 906
FACTORY-BUILT BARBECUE APPLIANCES

906.1 General. Factory-built barbecue appliances shall be of an approved type and shall be installed in accordance with the manufacturer's installation instructions, this chapter and Chapters 3, 5, 7, 8 and the *International Fuel Gas Code*.

SECTION 907
INCINERATORS AND CREMATORIES

907.1 General. Incinerators and crematories shall be listed and labeled in accordance with UL 791 and shall be installed in accordance with the manufacturer's installation instructions.

SECTION 908
COOLING TOWERS, EVAPORATIVE CONDENSERS AND FLUID COOLERS

908.1 General. A cooling tower used in conjunction with an air-conditioning appliance shall be installed in accordance with the manufacturer's installation instructions.

908.2 Access. Cooling towers, evaporative condensers and fluid coolers shall be provided with ready access.

908.3 Location. Cooling towers, evaporative condensers and fluid coolers shall be located to prevent the discharge vapor plumes from entering occupied spaces. Plume discharges shall be not less than 5 feet (1524 mm) above or 20 feet (6096 mm) away from any ventilation inlet to a building. Location on the property shall be as required for buildings in accordance with the *International Building Code*.

908.4 Support and anchorage. Supports for cooling towers, evaporative condensers and fluid coolers shall be designed in accordance with the *International Building Code*. Seismic restraints shall be as required by the *International Building Code*.

908.5 Water supply. Water supplies and protection shall be as required by the *International Plumbing Code*.

908.6 Drainage. Drains, overflows and blowdown provisions shall be indirectly connected to an approved disposal location. Discharge of chemical waste shall be approved by the appropriate regulatory authority.

908.7 Refrigerants and hazardous fluids. Heat exchange equipment that contains a refrigerant and that is part of a closed refrigeration system shall comply with Chapter 11. Heat exhange equipment containing heat transfer fluids which are flammable, combustible or hazardous shall comply with the *International Fire Code*.

SECTION 909
VENTED WALL FURNACES

909.1 General. Vented wall furnaces shall be installed in accordance with their listing and the manufacturer's installation instructions. Oil-fired furnaces shall be tested in accordance with UL 730.

909.2 Location. Vented wall furnaces shall be located so as not to cause a fire hazard to walls, floors, combustible furnishings or doors. Vented wall furnaces installed between bathrooms and adjoining rooms shall not circulate air from bathrooms to other parts of the building.

909.3 Door swing. Vented wall furnaces shall be located so that a door cannot swing within 12 inches (305 mm) of an air inlet or air outlet of such furnace measured at right angles to the opening. Doorstops or door closers shall not be installed to obtain this clearance.

909.4 Ducts prohibited. Ducts shall not be attached to wall furnaces. Casing extension boots shall not be installed unless listed as part of the appliance.

909.5 Manual shutoff valve. A manual shutoff valve shall be installed ahead of all controls.

909.6 Access. Vented wall furnaces shall be provided with access for cleaning of heating surfaces, removal of burners, replacement of sections, motors, controls, filters and other working parts, and for adjustments and lubrication of parts requiring such attention. Panels, grilles and access doors that must be removed for normal servicing operations shall not be attached to the building construction.

SECTION 910
FLOOR FURNACES

910.1 General. Floor furnaces shall be installed in accordance with their listing and the manufacturer's installation instructions. Oil-fired furnaces shall be tested in accordance with UL 729.

910.2 Placement. Floor furnaces shall not be installed in the floor of any aisle or passageway of any auditorium, public hall, place of assembly, or in any egress element from any such room or space.

With the exception of wall register models, a floor furnace shall not be placed closer than 6 inches (152 mm) to the nearest wall, and wall register models shall not be placed closer than 6 inches (152 mm) to a corner.

The furnace shall be placed such that a drapery or similar combustible object will not be nearer than 12 inches (305 mm) to any portion of the register of the furnace. Floor furnaces shall not be installed in concrete floor construction built on grade. The controlling thermostat for a floor furnace shall be located within the same room or space as the floor furnace or shall be located in an adjacent room or space that is permanently open to the room or space containing the floor furnace.

910.3 Bracing. The floor around the furnace shall be braced and headed with a support framework design in accordance with the *International Building Code*.

910.4 Clearance. The lowest portion of the floor furnace shall have not less than a 6-inch (152 mm) clearance from the grade level; except where the lower 6-inch (152 mm) portion of the floor furnace is sealed by the manufacturer to prevent entrance of water, the minimum clearance shall be reduced to not less than 2 inches (51 mm). Where these clearances are not present, the ground below and to the sides shall be excavated to form a pit under the furnace so that the required clearance is provided beneath the lowest portion of the furnace. A 12-inch (305 mm) minimum clearance shall be provided on all sides except the control side, which shall have an 18-inch (457 mm) minimum clearance.

SECTION 911
DUCT FURNACES

911.1 General. Duct furnaces shall be installed in accordance with the manufacturer's installation instructions. Electric furnaces shall be tested in accordance with UL 1995.

SECTION 912
INFRARED RADIANT HEATERS

912.1 Support. Infrared radiant heaters shall be safely and adequately fixed in an approved position independent of fuel and electric supply lines. Hangers and brackets shall be noncombustible material.

912.2 Clearances. Heaters shall be installed with clearances from combustible material in accordance with the manufacturer's installation instructions.

SECTION 913
CLOTHES DRYERS

913.1 General. Clothes dryers shall be installed in accordance with the manufacturer's installation instructions. Electric residential clothes dryers shall be tested in accordance with an approved test standard. Electric commercial clothes dryers shall be tested in accordance with UL 1240. Electric coin-operated clothes dryers shall be tested in accordance with UL 2158.

913.2 Exhaust required. Clothes dryers shall be exhausted in accordance with Section 504.

913.3 Clearances. Clothes dryers shall be installed with clearance to combustibles in accordance with the manufacturer's instructions.

SECTION 914
SAUNA HEATERS

914.1 Location and protection. Sauna heaters shall be located so as to minimize the possibility of accidental contact by a person in the room.

914.1.1 Guards. Sauna heaters shall be protected from accidental contact by an approved guard or barrier of material having a low coefficient of thermal conductivity. The guard shall not substantially affect the transfer of heat from the heater to the room.

914.2 Installation. Sauna heaters shall be listed and labeled and shall be installed in accordance with their listing and the manufacturer's installation instructions.

914.3 Access. Panels, grilles and access doors that are required to be removed for normal servicing operations shall not be attached to the building.

914.4 Heat and time controls. Sauna heaters shall be equipped with a thermostat that will limit room temperature to 194°F (90°C). If the thermostat is not an integral part of the sauna heater, the heat-sensing element shall be located within 6 inches (152 mm) of the ceiling. If the heat-sensing element is a capillary tube and bulb, the assembly shall be attached to the wall or other support, and shall be protected against physical damage.

914.4.1 Timers. A timer, if provided to control main burner operation, shall have a maximum operating time of 1 hour. The control for the timer shall be located outside the sauna room.

914.5 Sauna room. A ventilation opening into the sauna room shall be provided. The opening shall be not less than 4 inches by 8 inches (102 mm by 203 mm) located near the top of the door into the sauna room.

914.5.1 Warning notice. The following permanent notice, constructed of approved material, shall be mechanically attached to the sauna room on the outside:

WARNING: DO NOT EXCEED 30 MINUTES IN SAUNA. EXCESSIVE EXPOSURE CAN BE HARMFUL TO HEALTH. ANY PERSON WITH POOR HEALTH SHOULD CONSULT A PHYSICIAN BEFORE USING SAUNA.

The words shall contrast with the background and the wording shall be in letters not less than 0.25-inch (6.4 mm) high.

Exception: This section shall not apply to one- and two-family dwellings.

SECTION 915
ENGINE AND GAS TURBINE-POWERED EQUIPMENT AND APPLIANCES

915.1 General. The installation of liquid-fueled stationary internal combustion engines and gas turbines, including fuel storage and piping, shall meet the requirements of NFPA 37.

915.2 Powered equipment and appliances. Permanently installed equipment and appliances powered by internal combustion engines and turbines shall be installed in accordance with the manufacturer's installation instructions and NFPA 37.

SECTION 916
POOL AND SPA HEATERS

916.1 General. Pool and spa heaters shall be installed in accordance with the manufacturer's installation instructions. Oil-fired pool and spa heaters shall be tested in accordance with UL 726. Electric pool and spa heaters shall be tested in accordance with UL 1261.

SECTION 917
COOKING APPLIANCES

917.1 Cooking appliances. Cooking appliances that are designed for permanent installation, including ranges, ovens, stoves, broilers, grills, fryers, griddles and barbecues, shall be listed, labeled and installed in accordance with the manufacturer's installation instructions. Oil-burning stoves shall be tested in accordance with UL 896. Solid fuel-fired ovens shall be tested in accordance with UL 2162.

917.2 Prohibited location. Cooking appliances designed, tested, listed and labeled for use in commercial occupancies shall not be installed within dwelling units or within any area where domestic cooking operations occur.

917.3 Domestic appliances. Cooking appliances installed within dwelling units and within areas where domestic cooking operations occur shall be listed and labeled as household-type appliances for domestic use

SECTION 918
FORCED-AIR WARM-AIR FURNACES

918.1 Forced-air furnaces. Oil-fired furnaces shall be tested in accordance with UL 727. Electric furnaces shall be tested in accordance with UL 1995. Solid fuel furnaces shall be tested in accordance with UL 391. Forced-air furnaces shall be installed in accordance with the listings and the manufacturer's installation instructions.

918.2 Minimum duct sizes. The minimum unobstructed total area of the outside and return air ducts or openings to a forced-air warm-air furnace shall be not less than 2 square inches per 1,000 Btu/h (4402 mm²/kW) output rating capacity of the furnace and not less than that specified in the furnace manufacturer's installation instructions. The minimum unobstructed total area of supply ducts from a forced-air warm-air furnace shall be not less than 2 square inches for each 1,000 Btu/h (4402 mm²/kW) output rating capacity of the furnace and not less than that specified in the furnace manufacturer's installation instructions.

Exception: The total area of the supply air ducts and outside and return air ducts shall not be required to be larger than the minimum size required by the furnace manufacturer's installation instructions.

918.3 Heat pumps. The minimum unobstructed total area of the outside and return air ducts or openings to a heat pump shall be not less than 6 square inches per 1,000 Btu/h (13 208

mm²/kW) output rating or as indicated by the conditions of listing of the heat pump. Electric heat pumps shall be tested in accordance with UL 1995.

918.4 Dampers. Volume dampers shall not be placed in the air inlet to a furnace in a manner that will reduce the required air to the furnace.

918.5 Circulating air ducts for forced-air warm-air furnaces. Circulating air for fuel-burning, forced-air-type, warm-air furnaces shall be conducted into the blower housing from outside the furnace enclosure by continuous air-tight ducts.

918.6 Prohibited sources. Outside or return air for a forced-air heating system shall not be taken from the following locations:

1. Closer than 10 feet (3048 mm) from an appliance vent outlet, a vent opening from a plumbing drainage system or the discharge outlet of an exhaust fan, unless the outlet is 3 feet (914 mm) above the outside air inlet.

2. Where there is the presence of objectionable odors, fumes or flammable vapors; or where located less than 10 feet (3048 mm) above the surface of any abutting public way or driveway; or where located at grade level by a sidewalk, street, alley or driveway.

3. A hazardous or insanitary location or a refrigeration machinery room as defined in this code.

4. A room or space, the volume of which is less than 25 percent of the entire volume served by such system. Where connected by a permanent opening having an area sized in accordance with Sections 918.2 and 918.3, adjoining rooms or spaces shall be considered as a single room or space for the purpose of determining the volume of such rooms or spaces.

   Exception: The minimum volume requirement shall not apply where the amount of return air taken from a room or space is less than or equal to the amount of supply air delivered to such room or space.

5. A closet, bathroom, toilet room, kitchen, garage, mechanical room, boiler room or furnace room.

6. A room or space containing a fuel-burning appliance where such room or space serves as the sole source of return air.

   Exceptions:

   1. This shall not apply where the fuel-burning appliance is a direct-vent appliance.

   2. This shall not apply where the room or space complies with the following requirements:

      2.1. The return air shall be taken from a room or space having a volume exceeding 1 cubic foot for each 10 Btu/h (9.6 L/W) of combined input rating of all fuel-burning appliances therein.

      2.2. The volume of supply air discharged back into the same space shall be approximately equal to the volume of return air taken from the space.

      2.3. Return-air inlets shall not be located within 10 feet (3048 mm) of any appliance firebox or draft hood in the same room or space.

   3. This shall not apply to rooms or spaces containing solid fuel-burning appliances, provided that return-air inlets are located not less than 10 feet (3048 mm) from the firebox of such appliances.

918.7 Outside opening protection. Outdoor air intake openings shall be protected in accordance with Section 401.6.

918.8 Return-air limitation. Return air from one dwelling unit shall not be discharged into another dwelling unit.

SECTION 919
CONVERSION BURNERS

919.1 Conversion burners. The installation of conversion burners shall conform to ANSI Z21.8.

SECTION 920
UNIT HEATERS

920.1 General. Unit heaters shall be installed in accordance with the listing and the manufacturer's installation instructions. Oil-fired unit heaters shall be tested in accordance with UL 731.

920.2 Support. Suspended-type unit heaters shall be supported by elements that are designed and constructed to accommodate the weight and dynamic loads. Hangers and brackets shall be of noncombustible material. Suspended-type oil-fired unit heaters shall be installed in accordance with NFPA 31.

920.3 Ductwork. A unit heater shall not be attached to a warm-air duct system unless listed for such installation.

SECTION 921
VENTED ROOM HEATERS

921.1 General. Vented room heaters shall be listed and labeled and shall be installed in accordance with the conditions of the listing and the manufacturer's instructions.

SECTION 922
KEROSENE AND OIL-FIRED STOVES

922.1 General. Kerosene and oil-fired stoves shall be listed and labeled and shall be installed in accordance with the conditions of the listing and the manufacturer's installation instructions. Kerosene and oil-fired stoves shall comply with NFPA 31. Oil-fired stoves shall be tested in accordance with UL 896.

SECTION 923
SMALL CERAMIC KILNS

923.1 General. The provisions of this section shall apply to kilns that are used for ceramics, have a maximum interior volume of 20 cubic feet (0.566 m³) and are used for hobby and noncommercial purposes.

923.1.1 Installation. Kilns shall be installed in accordance with the manufacturer's installation instructions and the provisions of this code.

SECTION 924
STATIONARY FUEL CELL POWER PLANTS

924.1 General. Stationary fuel cell power plants having a power output not exceeding 1,000 kW, shall be tested in accordance with ANSI Z21.83 and shall be installed in accordance with the manufacturer's installation instructions and NFPA 853.

SECTION 925
MASONRY HEATERS

925.1 General. Masonry heaters shall be constructed in accordance with the *International Building* Code.

CHAPTER 10

BOILERS, WATER HEATERS AND PRESSURE VESSELS

SECTION 1001
GENERAL

1001.1 Scope. This chapter shall govern the installation, alteration and repair of boilers, water heaters and pressure vessels.

Exceptions:

1. Pressure vessels used for unheated water supply.

2. Portable unfired pressure vessels and Interstate Commerce Commission containers.

3. Containers for bulk oxygen and medical gas.

4. Unfired pressure vessels having a volume of 5 cubic feet (0.14 m³) or less operating at pressures not exceeding 250 pounds per square inch (psi) (1724 kPa) and located within occupancies of Groups B, F, H, M, R, S and U.

5. Pressure vessels used in refrigeration systems that are regulated by Chapter 11 of this code.

6. Pressure tanks used in conjunction with coaxial cables, telephone cables, power cables and other similar humidity control systems.

7. Any boiler or pressure vessel subject to inspection by federal or state inspectors.

SECTION 1002
WATER HEATERS

1002.1 General. Potable water heaters and hot water storage tanks shall be listed and labeled and installed in accordance with the manufacturer's installation instructions, the *International Plumbing Code* and this code. All water heaters shall be capable of being removed without first removing a permanent portion of the building structure. The potable water connections and relief valves for all water heaters shall conform to the requirements of the *International Plumbing Code*. Domestic electric water heaters shall comply with UL 174 or UL 1453. Commercial electric water heaters shall comply with UL 1453. Oil-fired water heaters shall comply with UL 732.

1002.2 Water heaters utilized for space heating. Water heaters utilized both to supply potable hot water and provide hot water for space-heating applications shall be listed and labeled for such applications by the manufacturer and shall be installed in accordance with the manufacturer's installation instructions and the *International Plumbing Code*.

1002.2.1 Sizing. Water heaters utilized for both potable water heating and space-heating applications shall be sized to prevent the space-heating load from diminishing the required potable water-heating capacity.

1002.2.2 Scald protection. Where a combination potable water-heating and space-heating system requires water for space heating at temperatures higher than 140°F (60°C), a tempering valve shall be provided to temper the water supplied to the potable hot water distribution system to a temperature of 140°F (60°C) or less.

1002.3 Supplemental water-heating devices. Potable water-heating devices that utilize refrigerant-to-water heat exchangers shall be approved and installed in accordance with the *International Plumbing Code* and the manufacturer's installation instructions.

SECTION 1003
PRESSURE VESSELS

1003.1 General. All pressure vessels shall bear the label of an approved agency and shall be installed in accordance with the manufacturer's installation instructions.

1003.2 Piping. All piping materials, fittings, joints, connections and devices associated with systems utilized in conjunction with pressure vessels shall be designed for the specific application and shall be approved.

1003.3 Welding. Welding on pressure vessels shall be performed by approved welders in compliance with nationally recognized standards.

SECTION 1004
BOILERS

1004.1 Standards. Oil-fired boilers and their control systems shall be listed and labeled in accordance with UL 726. Electric boilers and their control systems shall be listed and labeled in accordance with UL 834. Boilers shall be designed and constructed in accordance with the requirements of ASME CSD-1 and as applicable, the ASME *Boiler and Pressure Vessel Code*, Sections I, II, V, and IX; NFPA 8501; NFPA 8502 or NFPA 8504.

1004.2 Installation. In addition to the requirements of this code, the installation of boilers shall conform to the manufacturer's instructions. Operating instructions of a permanent type shall be attached to the boiler. Boilers shall have all controls set, adjusted and tested by the installer. The manufacturer's rating data and the nameplate shall be attached to the boiler.

1004.3 Working clearance. Clearances shall be maintained around boilers, generators, heaters, tanks and related equipment and appliances so as to permit inspection, servicing, repair, replacement and visibility of all gauges. When boilers are installed or replaced, clearance shall be provided to allow access for inspection, maintenance and repair. Passageways around all sides of boilers shall have an unobstructed width of not less than 18 inches (457 mm), unless otherwise approved.

1004.3.1 Top clearance. High-pressure steam boilers having a steam-generating capacity in excess of 5,000 pounds per hour (2268 kg/h) or having a heating surface in excess of 1,000 square feet (93 m²) or input in excess of 5,000,000 Btu/h (1465 kW) shall have a minimum clearance of 7 feet (2134 mm) from the top of the boiler to the ceiling. Steam-heating boilers and hot-water-heating boilers that exceed one of the following limits: 5,000,000 Btu/h input (1465 kW); 5,000 pounds of steam per hour (2268 kg/h) capacity or a 1,000-square-foot (93 m²) heating surface; and high-pressure steam boilers that do not exceed one of the following limits: 5,000,000 Btu/h input (1465 kW); 5,000 pounds of steam per hour (2268 kg/h) capacity or a 1,000-square-foot (93 m²) heating surface; and all boilers with manholes on top of the boiler, shall have a minimum clearance of 3 feet (914 mm) from the top of the boiler to the ceiling. Package boilers, steam-heating boilers and hot-water-heating boilers without manholes on top of the shell and not exceeding one of the limits of this section shall have a minimum clearance of 2 feet (610 mm) from the ceiling.

1004.4 Mounting. Equipment shall be set or mounted on a level base capable of supporting and distributing the weight contained thereon. Boilers, tanks and equipment shall be secured in accordance with the manufacturer's installation instructions.

1004.5 Floors. Boilers shall be mounted on floors of noncombustible construction, unless listed for mounting on combustible flooring.

1004.6 Boiler rooms and enclosures. Boiler rooms and enclosures and access thereto shall comply with the *International Building Code* and Chapter 3 of this code. Boiler rooms shall be equipped with a floor drain or other approved means for disposing of liquid waste.

1004.7 Operating adjustments and instructions. Hot water and steam boilers shall have all operating and safety controls set and operationally tested by the installing contractor. A complete control diagram and boiler operating instructions shall be furnished by the installer for each installation.

SECTION 1005
BOILER CONNECTIONS

1005.1 Valves. Every boiler or modular boiler shall have a shutoff valve in the supply and return piping. For multiple boiler or multiple modular boiler installations, each boiler or modular boiler shall have individual shutoff valves in the supply and return piping.

> **Exception:** Shutoff valves are not required in a system having a single low-pressure steam boiler.

1005.2 Potable water supply. The water supply to all boilers shall be connected in accordance with the *International Plumbing Code*.

SECTION 1006
SAFETY AND PRESSURE RELIEF VALVES
AND CONTROLS

1006.1 Safety valves for steam boilers. All steam boilers shall be protected with a safety valve.

1006.2 Safety relief valves for hot water boilers. Hot water boilers shall be protected with a safety relief valve.

1006.3 Pressure relief for pressure vessels. All pressure vessels shall be protected with a pressure relief valve or pressure-limiting device as required by the manufacturer's installation instructions for the pressure vessel.

1006.4 Approval of safety and safety relief valves. Safety and safety relief valves shall be listed and labeled, and shall have a minimum rated capacity for the equipment or appliances served. Safety and safety relief valves shall be set at a maximum of the nameplate pressure rating of the boiler or pressure vessel.

1006.5 Installation. Safety or relief valves shall be installed directly into the safety or relief valve opening on the boiler or pressure vessel. Valves shall not be located on either side of a safety or relief valve connection. The relief valve shall discharge by gravity.

1006.6 Safety and relief valve discharge. Safety and relief valve discharge pipes shall be of rigid pipe that is approved for the temperature of the system. The discharge pipe shall be the same diameter as the safety or relief valve outlet. Safety and relief valves shall not discharge so as to be a hazard, a potential cause of damage or otherwise a nuisance. High-pressure-steam safety valves shall be vented to the outside of the structure. Where a low-pressure safety valve or a relief valve discharges to the drainage system, the installation shall conform to the *International Plumbing Code*.

1006.7 Boiler safety devices. Boilers shall be equipped with controls and limit devices as required by the manufacturer's installation instructions and the conditions of the listing.

1006.8 Electrical requirements. The power supply to the electrical control system shall be from a two-wire branch circuit that has a grounded conductor, or from an isolation transformer with a two-wire secondary. Where an isolation transformer is provided, one conductor of the secondary winding shall be grounded. Control voltage shall not exceed 150 volts nominal, line to line. Control and limit devices shall interrupt the ungrounded side of the circuit. A means of manually disconnecting the control circuit shall be provided and controls shall be arranged so that when deenergized, the burner shall be inoperative. Such disconnecting means shall be capable of being locked in the off position and shall be provided with ready access.

SECTION 1007
BOILER LOW-WATER CUTOFF

1007.1 General. All steam and hot water boilers shall be protected with a low-water cutoff control.

1007.2 Operation. The low-water cutoff shall automatically stop the combustion operation of the appliance when the water level drops below the lowest safe water level as established by the manufacturer.

SECTION 1008
STEAM BLOWOFF VALVE

1008.1 General. Every steam boiler shall be equipped with a quick-opening blowoff valve. The valve shall be installed in the opening provided on the boiler. The minimum size of the valve shall be the size specified by the boiler manufacturer or the size of the boiler blowoff-valve opening.

1008.2 Discharge. Blowoff valves shall discharge to a safe place of disposal. Where discharging to the drainage system, the installation shall conform to the *International Plumbing Code*.

SECTION 1009
HOT WATER BOILER EXPANSION TANK

1009.1 Where required. An expansion tank shall be installed in every hot water system. For multiple boiler installations, a minimum of one expansion tank is required. Expansion tanks shall be of the closed or open type. Tanks shall be rated for the pressure of the hot water system.

1009.2 Closed-type expansion tanks. Closed-type expansion tanks shall be installed in accordance with the manufacturer's instructions. The size of the tank shall be based on the capacity of the hot-water-heating system. The minimum size of the tank shall be determined in accordance with the following equation:

$$V_t = \frac{(0.00041T - 0.0466)V_s}{\left(\frac{P_a}{P_f}\right) - \left(\frac{P_a}{P_o}\right)}$$ (Equation 10-1)

For SI:

$$V_t = \frac{(0.000738T - 0.03348)V_s}{\left(\frac{P_a}{P_f}\right) - \left(\frac{P_a}{P_o}\right)}$$

where:

V_t = Minimum volume of tanks (gallons) (L).

V_s = Volume of system, not including expansion tanks(gallons) (L).

T = Average operating temperature (°F) (°C).

P_a = Atmospheric pressure (psi) (kPa).

P_f = Fill pressure (psi) (kPa).

P_o = Maximum operating pressure (psi) (kPa).

1009.3 Open-type expansion tanks. Open-type expansion tanks shall be located a minimum of 4 feet (1219 mm) above the highest heating element. The tank shall be adequately sized for the hot water system. An overflow with a minimum diameter of 1 inch (25 mm) shall be installed at the top of the tank. The overflow shall discharge to the drainage system in accordance with the *International Plumbing Code*.

SECTION 1010
GAUGES

1010.1 Hot water boiler gauges. Every hot water boiler shall have a pressure gauge and a temperature gauge, or a combination pressure and temperature gauge. The gauges shall indicate the temperature and pressure within the normal range of the system's operation.

1010.2 Steam boiler gauges. Every steam boiler shall have a water-gauge glass and a pressure gauge. The pressure gauge shall indicate the pressure within the normal range of the system's operation.

1010.2.1 Water-gauge glass. The gauge glass shall be installed so that the midpoint is at the normal boiler water level.

SECTION 1011
TESTS

1011.1 Tests. Upon completion of the assembly and installation of boilers and pressure vessels, acceptance tests shall be conducted in accordance with the requirements of the ASME *Boiler and Pressure Vessel Code*. Where field assembly of pressure vessels or boilers is required, a copy of the completed U-1 Manufacturer's Data Report required by the ASME *Boiler and Pressure Vessel Code* shall be submitted to the code official.

1011.2 Test gauges. An indicating test gauge shall be connected directly to the boiler or pressure vessel where it is visible to the operator throughout the duration of the test. The pressure gauge scale shall be graduated over a range of not less than one and one-half times and not greater than four times the maximum test pressure. All gauges utilized for testing shall be calibrated and certified by the test operator.

CHAPTER 11

REFRIGERATION

SECTION 1101
GENERAL

1101.1 Scope. This chapter shall govern the design, installation, construction and repair of refrigeration systems that vaporize and liquefy a fluid during the refrigerating cycle. Refrigerant piping design and installation, including pressure vessels and pressure relief devices, shall conform to this code. Permanently installed refrigerant storage systems and other components shall be considered as part of the refrigeration system to which they are attached.

1101.2 Factory-built equipment and appliances. Listed and labeled self-contained, factory-built equipment and appliances shall be tested in accordance with UL 207, 412, 471 or 1995. Such equipment and appliances are deemed to meet the design, manufacture and factory test requirements of this code if installed in accordance with their listing and the manufacturer's installation instructions.

1101.3 Protection. Any portion of a refrigeration system that is subject to physical damage shall be protected in an approved manner.

1101.4 Water connection. Water supply and discharge connections associated with refrigeration systems shall be made in accordance with this code and the *International Plumbing Code*.

1101.5 Fuel gas connection. Fuel gas devices, equipment and appliances used with refrigeration systems shall be installed in accordance with the *International Fuel Gas Code*.

1101.6 General. Refrigeration systems shall comply with the requirements of this code and, except as modified by this code, ASHRAE 15. Ammonia-refrigerating systems shall comply with this code and, except as modified by this code, ASHRAE 15 and IIAR 2.

1101.7 Maintenance. Mechanical refrigeration systems shall be maintained in proper operating condition, free from accumulations of oil, dirt, waste, excessive corrosion, other debris and leaks.

1101.8 Change in refrigerant type. The type of refrigerant in refrigeration systems having a refrigerant circuit containing more than 220 pounds of Group A1 or 30 pounds of any other group refrigerant shall not be changed without prior notification to the code official and compliance with the applicable code provisions for the new refrigerant type.

[F] 1101.9 Refrigerant discharge. Notification of refrigerant discharge shall be provided in accordance with the *International Fire Code*.

SECTION 1102
SYSTEM REQUIREMENTS

1102.1 General. The system classification, allowable refrigerants, maximum quantity, enclosure requirements, location limitations, and field pressure test requirements shall be determined as follows:

1. Determine the refrigeration system's classification, in accordance with Section 1103.3.

2. Determine the refrigerant classification in accordance with Section 1103.1.

3. Determine the maximum allowable quantity of refrigerant in accordance with Section 1104, based on type of refrigerant, system classification, and occupancy.

4. Determine the system enclosure requirements in accordance with Section 1104.

5. Refrigeration equipment and appliance location and installation shall be subject to the limitations of Chapter 3.

6. Nonfactory-tested, field-erected equipment and appliances shall be pressure tested in accordance with Section 1108.

1102.2 Refrigerants. The refrigerant shall be that which the equipment or appliance was designed to utilize or converted to utilize. Refrigerants not identified in Table 1103.1 shall be approved before use.

1102.2.1 Mixing. Refrigerants, including refrigerant blends, with different designations in ASHRAE 34 shall not be mixed in a system.

Exception: Addition of a second refrigerant is allowed where permitted by the equipment or appliance manufacturer to improve oil return at low temperatures. The refrigerant and amount added shall be in accordance with the manufacturer's instructions.

1102.2.2 Purity. Refrigerants used in refrigeration systems shall be new, recovered or reclaimed refrigerants in accordance with Section 1102.2.2.1, 1102.2.2.2 or 1102.2.2.3. Where required by the equipment or appliance owner or the code official, the installer shall furnish a signed declaration that the refrigerant used meets the requirements of Section 1102.2.2.1, 1102.2.2.2 or 1102.2.2.3.

Exception: The refrigerant used shall meet the purity specifications set by the manufacturer of the equipment or appliance in which such refrigerant is used where such specifications are different from that specified in Sections 1102.2.2.1, 1102.2.2.2 and 1102.2.2.3.

1102.2.2.1 New refrigerants. Refrigerants shall be of a purity level specified by the equipment or appliance manufacturer.

1102.2.2.2 Recovered refrigerants. Refrigerants that are recovered from refrigeration and air-conditioning systems shall not be reused in other than the system from which they were recovered and in other systems of the same owner. Recovered refrigerants shall be filtered and dried before reuse. Recovered refrigerants that show clear signs of contamination shall not be reused unless reclaimed in accordance with Section 1102.2.2.3.

1102.2.2.3 Reclaimed refrigerants. Used refrigerants shall not be reused in a different owner's equipment or appliances unless tested and found to meet the purity requirements of ARI 700. Contaminated refrigerants shall not be used unless reclaimed and found to meet the purity requirements of ARI 700.

SECTION 1103
REFRIGERATION SYSTEM CLASSIFICATION

1103.1 Refrigerant classification. Refrigerants shall be classified in accordance with ASHRAE 34 as listed in Table 1103.1.

1103.2 Occupancy classification. Locations of refrigerating systems are described by occupancy classifications that consider the ability of people to respond to potential exposure to refrigerants. Where equipment or appliances, other than piping, are located outside a building and within 20 feet (6096 mm) of any building opening, such equipment or appliances shall be governed by the occupancy classification of the building. Occupancy classifications shall be defined as follows:

1. Institutional occupancy is that portion of premises from which, because they are disabled, debilitated or confined, occupants cannot readily leave without the assistance of others. Institutional occupancies include, among others, hospitals, nursing homes, asylums and spaces containing locked cells.

2. Public assembly occupancy is that portion of premises where large numbers of people congregate and from which occupants cannot quickly vacate the space. Public assembly occupancies include, among others, auditoriums, ballrooms, classrooms, passenger depots, restaurants and theaters.

3. Residential occupancy is that portion of premises that provides the occupants with complete independent living facilities, including permanent provisions for living, sleeping, eating, cooking and sanitation. Residential occupancies include, among others, dormitories, hotels, multiunit apartments and private residences.

4. Commercial occupancy is that portion of premises where people transact business, receive personal service or purchase food and other goods. Commercial occupancies include, among others, office and professional buildings, markets (but not large mercantile occupancies) and work or storage areas that do not qualify as industrial occupancies.

5. Large mercantile occupancy is that portion of premises where more than 100 persons congregate on levels above or below street level to purchase personal merchandise.

6. Industrial occupancy is that portion of premises that is not open to the public, where access by authorized persons is controlled, and that is used to manufacture, process or store goods such as chemicals, food, ice, meat or petroleum.

7. Mixed occupancy occurs when two or more occupancies are located within the same building. When each occupancy is isolated from the rest of the building by tight walls, floors and ceilings and by self-closing doors, the requirements for each occupancy shall apply to its portion of the building. When the various occupancies are not so isolated, the occupancy having the most stringent requirements shall be the governing occupancy.

1103.3 System classification. Refrigeration systems shall be classified according to the degree of probability that refrigerant leaked from a failed connection, seal, or component could enter an occupied area. The distinction is based on the basic design or location of the components.

1103.3.1 Low-probability systems. Double-indirect open-spray systems, indirect closed systems and indirect-vented closed systems shall be classified as low-probability systems, provided that all refrigerant-containing piping and fittings are isolated when the quantities in Table 1103.1 are exceeded.

1103.3.2 High-probability systems. Direct systems and indirect open-spray systems shall be classified as high-probability systems.

Exception: An indirect open-spray system shall not be required to be classified as a high-probability system if the pressure of the secondary coolant is at all times (operating and standby) greater than the pressure of the refrigerant.

SECTION 1104
SYSTEM APPLICATION REQUIREMENTS

1104.1 General. The refrigerant, occupancy and system classification cited in this section shall be determined in accordance with Sections 1103.1, 1103.2 and 1103.3, respectively. For refrigerant blends assigned dual classifications, as formulated and for the worst case of fractionation, the classifications for the worst case of fractionation shall be used.

1104.2 Machinery room. Except as provided in Sections 1104.2.1 and 1104.2.2, all components containing the refrigerant shall be located either outdoors or in a machinery room where the quantity of refrigerant in an independent circuit of a system exceeds the amounts shown in Table 1103.1. For refrigerant blends not listed in Table 1103.1, the same requirement shall apply when the amount for any blend component exceeds that indicated in Table 1103.1 for that component. This requirement shall also apply when the combined amount of the blend components exceeds a limit of 69,100 parts per million (ppm) by volume. Machinery rooms required by this section shall be constructed and maintained in accordance with Section 1105 for Group A1 and B1 refrigerants and in accordance with Sections 1105 and 1106 for Group A2, B2, A3 and B3 refrigerants.

Exceptions:

1. Machinery rooms are not required for listed equipment and appliances containing not more than 6.6 pounds (3 kg) of refrigerant, regardless of the refrigerant's safety classification, where installed in accordance with the equipment's or appliance's listing and the equipment or appliance manufacturer's installation instructions.

[F] TABLE 1103.1
REFRIGERANT CLASSIFICATION, AMOUNT AND TLV-TWA

REFRIGERANT	CHEMICAL FORMULA	CHEMICAL NAME OR BLEND	HAZARD CATEGORIES[a]	REFRIGERANT CLASSIFICATION	DEGREES OF HAZARD[b]	[M] AMOUNT OF REFRIGERANT PER OCCUPIED SPACE			TLV-TWA[f] (ppm)
						Pounds per 1,000 cubic feet	ppm	g/m	
R-11[e]	CCl_3F	Trichlorofluoromethane	OHH	A1	2-0-0[c]	0.39	1,100	6.2	C1,000
R-12[e]	CCl_2F_2	Dichlorodifluoromethane	CG,OHH	A1	2-0-0[c]	5.6	18,000	90	1,000
R-13[e]	$CClF_3$	Chlorotrifluoromethane	CG,OHH	A1	2-0-0[c]	18	67,000	290	1,000
R-13B1[e]	$CBrF_3$	Bromotrifluoromethane	CG,OHH	A1	2-0-0[c]	22	57,000	350	1,000
R-14	CF_4	Tetrafluoromethane (carbon tetrafluoride)	CG,OHH	A1	2-0-0[c]	16	69,000	250	1,000
R-22	$CHClF_2$	Chlorodifluoromethane	CG,OHH	A1	2-0-0[c]	5.5	25,000	89	1,000
R-23	CHF_3	Trifluoromethane (fluoroform)	CG,OHH	A1	2-0-0[c]	7.3	41,000	120	1,000
R-32	CH_2, F_2	Difluoromethane (methylene fluoride)	CG,F,OHH	A2		4.2	32,000	68	—
R-113[e]	CCl_3FCClF_2	1, 1,2-trichloro-1,2,2-trifluoroethane	OHH	A1	2-0-0[c]	1.2	2,600	20	1,000
R-114[e]	$CClF_2CClF_2$	1,2-dichloro-1,2,2-tetrafluoroethane	CG,OHH	A1	2-0-0[c]	8.7	20,000	140	1,000
R-116	CF_3CF_3	Hexafluoroethane	CH,OHH	A1	1-0-0	24	69,000	390	—
R-123	$CHCl_2CF_3$	2,2-dichloro-1,1,1-trifluoroethane	OHH	B1	2-0-0[c]	3.5	9,100	57	50
R-124	$CHClFCF_3$	2-chloro-1,1,1,2-tetrafluoroethane	CG,OHH	A1	2-0-0[c]	3.5	10,000	56	1,000
R-125	CHF_2CF_3	Pentafluoroethane	CG,OHH	A1	2-0-0[c]	21	69,000	340	—
R-134a	CH_2FCF_3	1,1,1,2-tetrafluoroethane	CG,OHH	A1	2-0-0[c]	13	50,000	210	1,000
R-143a	CH_3CF_3	1,1,1-trifluoroethane	CG,F,OHH	A2	2-0-0[c]	3.8	18,000	60	—
R-152a	CH_3CHF_2	1,1-difluoroethane	CG,F,OHH	A2	1-4-0	1.6	9,300	25	—
R-170	CH_3CH_3	Ethane	CG,F,OHH	A3	2-4-0	0.54	7,000	8.7	1,000
R-218	$CF_3CF_2CF_3$	Octafluoropropane	CG,OHH	A1	2-0-0[c]	33	69,000	530	—
R-236fa	CF_3,CH_2CF_3	1,1,1,3,3,3-hexafluoropropane	CG,OHH	A1	2-0-0[c]	21	55,000	—	1,000
R-245fa	CHF_2,CH_2, CF_3	1,1,1,3,3-pentafluoropropane	CG,OHH	B1	2-0-0[c]	12	34,000	—	300
R-290	$CH_3CH_2CH_3$	Propane	CG,F,OHH	A3	2-4-0	0.56	5,000	9.0	2,500
R400[e]	zeotrope	R-12/114	CG,OHH	A1	2-0-0[c]	9.3	26,000	150	—
R-401A	zeotrope	R-22/152a/124 (53/13/34)	CG,OHH	A1	2-0-0[c]	4.8	20,000	77	—
R-401B	zeotrope	R-22/152a/124 (61/11/28)	CG,OHH	A1	2-0-0[c]	4.9	21,000	79	—
R-401C	zeotrope	R-22/152a/124 (33/15/52)	CG,OHH	A1	2-0-0[c]	4.4	17,000	71	—
R-402A	zeotrope	R-125/290/22 (60/2/38)	CG,OHH	A1	2-0-0[c]	10	39,000	160	—
R-402B	zeotrope	R-125/290/22 (38/2/60)	CG,OHH	A1	2-0-0[c]	7.8	32,000	120	—
R-403A	zeotrope	R-290/22/218 (5/75/20)	CG,OHH	A1	2-0-0[c]	—	—	—	—
R-403B	zeotrope	R-290/22/218 (5/56/39)	CG,OHH	A1	2-0-0[c]	—	—	—	—
R-404A	zeotrope	R-125/143a/134a (44/52/4)	CG,OHH	A1	2-0-0[c]	17	69,000	280	—

(continued)

[F] TABLE 1103.1 continued
REFRIGERANT CLASSIFICATION, AMOUNT AND TLV-TWA

REFRIGERANT	CHEMICAL FORMULA	CHEMICAL NAME OR BLEND	HAZARD CATEGORIES[a]	REFRIGERANT CLASSIFICATION[b]	DEGREES OF HAZARD[b]	[M] AMOUNT OF REFRIGERANT PER OCCUPIED SPACE			TLV-TWA[f] (ppm)
						Pounds per 1,000 cubic feet	ppm	g/m	
R-406A	zeotrope	R-22/600a/142b (55/4/41)	CG,F,OHH	A2	—	—	—	—	—
R-407A	zeotrope	R-32/125/134a (20/40/40)	CG,OHH	A1	2-0-0[c]	16	69,000	260	—
R-407B	zeotrope	R-32/125/134a (10/70/20)	CG,OHH	A1	2-0-0[c]	18	69,000	290	—
R-407C	zeotrope	R-32/125/134a (23/25/52)	CG,OHH	A1	2-0-0[c]	15	69,000	240	—
R-407D	zeotrope	R-32/125/134a (15/15/70)	CG,OHH	A1	2-0-0[c]	15	65,000	240	—
R-407E	zeotrope	R-32/125/134a (25/15/60)	CG,OHH	A1	2-0-0[c]	15	69,000	240	—
R-408A	zeotrope	R-125/143a/22 (7/46/47)	CG,OHH	A1	2-0-0[c]	10	47,000	170	—
R-409A	zeotrope	R-22/124/142b (60/25/15)	CG,OHH	A1	2-0-0[c]	4.9	20,000	79	—
R-409B	zeotrope	R-22/124/142b (65/25/10)	CG,OHH	A1	2-0-0[c]	4.9	20,000	78	—
R-410A	zeotrope	R-32/125 (50/50)	CG,OHH	A1	2-0-0[c]	10	55,000	160	—
R-410B	zeotrope	R-32/125 (45/55)	CG,OHH	A1	2-0-0[c]	11	58,000	180	—
R-411A	zeotrope	R-1270/22/152a (1.5/87.5/11.0)	CG,F,OHH	A2	—	—	—	—	—
R-411B	zeotrope	R-1270/22/152a (3/94/3)	CG,F,OHH	A2	—	—	—	—	—
R-412A	zeotrope	R-22/318/142b (70/5/25)	CG,F,OHH	A2	—	—	—	—	—
R-413A	zeotrope	R-218/134a/600a (9/88/3)	CG,F,OHH	A2	—	—	—	—	—
R-414A	zeotrope	R-22/124/600a/142b (51/28.5/4/16.5)	CG,OHH	A1	—	—	—	—	—
R-414B	zeotrope	R-22/124/600a/142b (50/39/1.5/9.5)	CG,OHH	A1	—	—	—	—	—
R-416A	zeotrope	R-134a/124/600 (59/39.5/1.5)	CG,OHH	A1	2-0-0[c]	6	21,000	96	—
R-417A	zeotrope	R-125/134a/600 (45.5/50/3.5)	CG,OHH	A1	2-0-0[c]	—	—	—	—
R-500[c]	azeotrope	R-12/152a (73.8/26.2)	CG,OHH	A1	2-0-0[c]	7.4	29,000	120	1,000
R-502[c]	azeotrope	R-22/115 (48.8/51.2)	CG,OHH	A1	2-0-0[c]	10	35,000	160	1,000
R-503[c]	azeotrope	R-23/13 (40.1/59.9)	CG,OHH	A1	2-0-0[c]	15	67,000	240	1,000
R-507A	azeotrope	R-125/143a (50/50)	CG,OHH	A1	2-0-0[c]	17	69,000	280	—
R-508A	azeotrope	R-23/116 (39/61)	CG,OHH	A1	2-0-0[c]	14	55,000	220	—
R-508B	azeotrope	R-23/116 (46/54)	CG,OHH	A1	2-0-0[c]	13	52,000	200	—
R-509A	zeotrope	R-22/218 (44/56)	CG,OHH	A1	2-0-0[c]	12	38,000	190	—
R-600	$CH_3CH_2CH_2CH_3$	Butane	CG,F,OHH	A3	1-4-0	—	—	—	—
R-600a	$CH(CH_3)2-CH_3$	Isobutane (2-methyl propane)	CG,F,OHH	A3	2-4-0	0.51	2,500	6.0	800

(continued)

[F] TABLE 1103.1 (continued)
REFRIGERANT CLASSIFICATION, AMOUNT AND TLV-TWA

REFRIGERANT	CHEMICAL FORMULA	CHEMICAL NAME OR BLEND	HAZARD CATEGORIES[a]	REFRIGERANT CLASSIFICATION	DEGREES OF HAZARD[b]	[M] AMOUNT OF REFRIGERANT PER OCCUPIED SPACE			TLV-TWA[f] (ppm)
						Pounds per 1,000 cubic feet	ppm	g/m	
R-717	NH_3	Ammonia	CG,C,F,OHH	B2	3-3-0[d]	0.022	500	0.35	25
R-718	H_2O	Water	—	A1	0-0-0	—	—	—	—
R-744	CO_2	Carbon dioxide	CG,OHH	A1	2-0-0[c]	4.5	40,000	72	5,000
R-1150	$CH_2=CH_2$	Ethene (ethylene)	CG,F,OHH	A3	1-4-2	0.38	5,200	6.0	1,000
R-1270	$CH_3CH=CH_2$	Propene (propylene)	CG,F,OHH	B3	1-4-1	0.37	3,400	5.0	660

For SI: 1 pound = 0.454 kg, 1 cubic foot = 0.0283 m^3

a. CG = Compressed gas; C = Corrosive; F = Flammable; OHH = Other Health Hazard.

b. Degrees of hazard are for health, fire, and reactivity, respectively, in accordance with NFPA 704.

c. Reduction to 1-0-0 is allowed if analysis satisfactory to the code official shows that the maximum concentration for a rupture or full loss of refrigerant charge would not exceed the IDLH, considering both the refrigerant quantity and room volume.

d. For installations that are entirely outdoors, use 3-1-0.

e. Class I ozone depleting substance; prohibited for new installations.

f. PEL or consistent occupational exposure limit on a time-weighted average (TWA) basis (unless noted C for ceiling) for an 8 hr/d and 40 hr/wk.

2. Piping in conformance with Section 1107 is allowed in other locations to connect components installed in a machinery room with those installed outdoors.

1104.2.1 Institutional occupancies. The amounts shown in Table 1103.1 shall be reduced by 50 percent for all areas of institutional occupancies except kitchens, laboratories, and mortuaries. The total of all Group A2, B2, A3 and B3 refrigerants shall not exceed 550 pounds (250 kg) in occupied areas or machinery rooms.

1104.2.2 Industrial occupancies and refrigerated rooms. This section applies only to industrial occupancies and refrigerated rooms for manufacturing, food and beverage preparation, meat cutting, other processes and storage. Machinery rooms are not required where all of the following conditions are met:

1. The space containing the machinery is separated from other occupancies by tight construction with tight-fitting doors.

2. Access is restricted to authorized personnel.

3. The floor area per occupant is not less than 100 square feet (9.3 m²) where machinery is located on floor levels with exits more than 6.6 feet (2012 mm) above the ground. Where provided with egress directly to the outdoors or into approved building exits, the minimum floor area shall not apply.

4. Refrigerant detectors are installed as required for machinery rooms in accordance with Section 1105.3.

5. Surfaces having temperatures exceeding 800°F (427°C) and open flames are not present where any Group A2, B2, A3 or B3 refrigerant is used (see Section 1104.3.4).

6. All electrical equipment and appliances conform to Class 1, Division 2, hazardous location classification requirements of NFPA 70 where the quantity of any Group A2, B2, A3 or B3 refrigerant in a single independent circuit would exceed 25 percent of the lower flammability limit (LFL) upon release to the space.

7. All refrigerant-containing parts in systems exceeding 100 hp (74.6 kW) drive power, except evaporators used for refrigeration or dehumidification; condensers used for heating; control and pressure relief valves for either;

and connecting piping, shall be located either outdoors or in a machinery room.

1104.3 Refrigerant restrictions. Refrigerant applications, maximum quantities and use shall be restricted in accordance with Sections 1104.3.1 through 1104.3.4.

1104.3.1 Air-conditioning for human comfort. In other than industrial occupancies where the quantity in a single independent circuit does not exceed the amount in Table 1103.1, Group B1, B2 and B3 refrigerants shall not be used in high-probability systems for air-conditioning for human comfort.

1104.3.2 Nonindustrial occupancies. Group A2 and B2 refrigerants shall not be used in high-probability systems where the quantity of refrigerant in any independent refrigerant circuit exceeds the amount shown in Table 1104.3.2. Group A3 and B3 refrigerants shall not be used except where approved.

Exception: This section does not apply to laboratories where the floor area per occupant is not less than 100 square feet (9.3 m²).

1104.3.3 All occupancies. The total of all Group A2, B2, A3 and B3 refrigerants other than R-717, ammonia, shall not exceed 1,100 pounds (499 kg) except where approved.

1104.3.4 Protection from refrigerant decomposition. Where any device having an open flame or surface temperature greater than 800°F (427°C) is used in a room containing more than 6.6 pounds (3 kg) of refrigerant in a single independent circuit, a hood and exhaust system shall be provided in accordance with Section 510. Such exhaust system shall exhaust combustion products to the outdoors.

Exception: A hood and exhaust system shall not be required:

1. Where the refrigerant is R-717, R-718, or R-744;

2. Where the combustion air is ducted from the outdoors in a manner that prevents leaked refrigerant from being combusted; or

3. Where a refrigerant detector is used to stop the combustion in the event of a refrigerant leak (see Sections 1105.3 and 1105.5).

1104.4 Volume calculations. Volume calculations shall be in accordance with Sections 1104.4.1 through 1104.4.3.

TABLE 1104.3.2
MAXIMUM PERMISSIBLE QUANTITIES OF REFRIGERANTS

TYPE OF REFRIGERATION SYSTEM	MAXIMUM POUNDS (kg) FOR VARIOUS OCCUPANCIES			
	Institutional	Assembly	Residential	All other occupancies
Sealed absorption system				
In exit access	0(0)	0(0)	3.3(1.5)	3.3(1.5)
In adjacent outdoor locations	0(0)	0(0)	22(10)	22(10)
In other than exit access	0(0)	6.6(3)	6.6(3)	6.6(3)
Unit systems				
In other than exit access	0(0)	0(0)	6.6(3)	6.6(3)

For SI: 1 pound = 0.454 kg.

1104.4.1 Noncommunicating spaces. Where the refrigerant-containing parts of a system are located in one or more spaces that do not communicate through permanent openings or HVAC ducts, the volume of the smallest, enclosed occupied space shall be used to determine the permissible quantity of refrigerant in the system.

1104.4.2 Communicating spaces. Where an evaporator or condenser is located in an air duct system, the volume of the smallest, enclosed occupied space served by the duct system shall be used to determine the maximum allowable quantity of refrigerant in the system.

> **Exception:** If airflow to any enclosed space cannot be reduced below one-quarter of its maximum, the entire space served by the air duct system shall be used to determine the maximum allowable quantity of refrigerant in the system.

1104.4.3 Plenums. Where the space above a suspended ceiling is continuous and part of the supply or return air plenum system, this space shall be included in calculating the volume of the enclosed space.

SECTION 1105
MACHINERY ROOM, GENERAL REQUIREMENTS

[B] 1105.1 Design and construction. Machinery rooms shall be designed and constructed in accordance with the *International Building Code* and this section.

1105.2 Openings. Ducts and air handlers in the machinery room that operate at a lower pressure than the room shall be sealed to prevent any refrigerant leakage from entering the airstream.

[F] 1105.3 Refrigerant detector. Refrigerant detectors in machinery rooms shall be provided as required by the *International Fire Code*.

1105.4 Tests. Periodic tests of the mechanical ventilating system shall be performed in accordance with manufacturer's specifications and as required by the code official.

1105.5 Fuel-burning appliances. Open flames that use combustion air from the machinery room shall not be installed in a machinery room.

> **Exceptions:**
> 1. Matches, lighters, halide leak detectors and similar devices.
> 2. Where the refrigerant is carbon dioxide or water.
> 3. Fuel-burning appliances shall not be prohibited in the same machinery room with refrigerant-containing equipment or appliances where combustion air is ducted from outside the machinery room and sealed in such a manner as to prevent any refrigerant leakage from entering the combustion chamber, or where a refrigerant vapor detector is employed to automatically shut off the combustion process in the event of refrigerant leakage.

1105.6 Ventilation. Machinery rooms shall be mechanically ventilated to the outdoors. Mechanical ventilation shall be capable of exhausting the minimum quantity of air both at normal operating and emergency conditions. Multiple fans or multispeed fans shall be allowed in order to produce the emergency ventilation rate and to obtain a reduced airflow for normal ventilation.

> **Exception:** Where a refrigerating system is located outdoors more than 20 feet (6096 mm) from any building opening and is enclosed by a penthouse, lean-to or other open structure, natural or mechanical ventilation shall be provided. Location of the openings shall be based on the relative density of the refrigerant to air. The free-aperture cross section for the ventilation of the machinery room shall be not less than:

$$F = \sqrt{G} \qquad \text{(Equation 11-1)}$$

For SI: $F = 0.138 \sqrt{G}$

where:

F = The free opening area in square feet (m²).

G = The mass of refrigerant in pounds (kg) in the largest system, any part of which is located in the machinery room.

1105.6.1 Discharge location. The discharge of the air shall be to the outdoors in accordance with Chapter 5. Exhaust from mechanical ventilation systems shall be discharged not less than 20 feet (6096 mm) from a property line or openings into buildings.

1105.6.2 Make-up air. Provisions shall be made for make-up air to replace that being exhausted. Openings for make-up air shall be located to avoid intake of exhaust air. Supply and exhaust ducts to the machinery room shall serve no other area, shall be constructed in accordance with Chapter 5 and shall be covered with corrosion-resistant screen of not less than $^1/_4$-inch (6.4 mm) mesh.

1105.6.3 Quantity—normal ventilation. During occupied conditions, the mechanical ventilation system shall exhaust the larger of the following:

1. Not less than 0.5 cfm per square foot (0.0025 m³/s m²) of machinery room area or 20 cfm (0.009 m³/s) per person; or
2. A volume required to limit the room temperature rise to 18°F (10°C) taking into account the ambient heating effect of all machinery in the room.

1105.6.4 Quantity—emergency conditions. Upon actuation of the refrigerant detector required in Section 1105.3, the mechanical ventilation system shall exhaust air from the machinery room in the following quantity:

$$Q = 100 \sqrt{G} \qquad \text{(Equation 11-2)}$$

For SI: $Q = 0.07 \sqrt{G}$

where:

Q = The airflow in cubic feet per minute (m³/s).

G = The design mass of refrigerant in pounds (kg) in the largest system, any part of which is located in the machinery room.

1105.7 Termination of relief devices. Pressure relief devices, fusible plugs and purge systems located within the machinery room shall terminate outside of the structure at a location not less than 15 feet (4572 mm) above the adjoining grade level and not less than 20 feet (6096 mm) from any window, ventilation opening or exit.

1105.8 Ammonia discharge. Pressure relief valves for ammonia systems shall discharge in accordance with ASHRAE 15.

SECTION 1106
MACHINERY ROOM, SPECIAL REQUIREMENTS

1106.1 General. Where required by Section 1104.2, the machinery room shall meet the requirements of this section in addition to the requirements of Section 1105.

1106.2 Elevated temperature. There shall not be an open flame-producing device or continuously operating hot surface over 800°F (427°C) permanently installed in the room.

1106.3 Ammonia room ventilation. Ventilation systems in ammonia machinery rooms shall be operated continuously.

Exceptions:

1. Machinery rooms equipped with a vapor detector that will automatically start the ventilation system and actuate an alarm at a detection level not to exceed 1,000 ppm; or

2. Machinery rooms conforming to the Class 1, Division 2, hazardous location classification requirements of NFPA 70.

1106.4 Flammable refrigerants. Where refrigerants of Groups A2, A3, B2 and B3 are used, the machinery room shall conform to the Class 1, Division 2, hazardous location classification requirements of NFPA 70.

Exception: Ammonia machinery rooms.

[F] 1106.5 Remote controls. Remote control of the mechanical equipment and appliances located in the machinery room shall be provided as required by the *International Fire Code.*

[F 1106.5.1 Refrigeration system. A clearly identified switch of the break-glass type shall provide off-only control of all electrically energized equipment and appliances in the machinery room, other than refrigerant leak detectors and machinery room ventilation.

[F 1106.5.2 Ventilation system. A clearly identified switch of the break-glass type shall provide on-only control of the machinery room ventilation fans.

[F]1106.6 Emergency signs and labels. Refrigeration units and systems shall be provided with approved emergency signs, charts, and labels in accordance with the *International Fire Code.*

SECTION 1107
REFRIGERANT PIPING

1107.1 General. All refrigerant piping shall be installed, tested and placed in operation in accordance with this chapter.

1107.2 Pipe enclosures. Rigid or flexible metal enclosures or pipe ducts shall be provided for soft, annealed copper tubing and used for refrigerant piping erected on the premises and containing other than Group A1 or B1 refrigerants. Enclosures shall not be required for connections between condensing units and the nearest riser box(es), provided such connections do not exceed 6 feet (1829 mm) in length.

1107.3 Condensation. All refrigerating piping and fittings, brine piping and fittings that, during normal operation, will reach a surface temperature below the dew point of the surrounding air, and are located in spaces or areas where condensation will cause a safety hazard to the building occupants, structure, electrical equipment or any other equipment or appliances, shall be protected in an approved manner to prevent such damage.

1107.4 Materials for refrigerant pipe and tubing. Piping materials shall be as set forth in Sections 1107.4.1 through 1107.4.5.

    1107.4.1 Steel pipe. Carbon steel pipe with a wall thickness not less than Schedule 80 shall be used for Group A2, A3, B2 or B3 refrigerant liquid lines for sizes 1.5 inches (38 mm) and smaller. Carbon steel pipe with a wall thickness not less than Schedule 40 shall be used for Group A1 or B1 refrigerant liquid lines 6 inches (152 mm) and smaller, Group A2, A3, B2 or B3 refrigerant liquid lines sizes 2 inches (51 mm) through 6 inches (152 mm) and all refrigerant suction and discharge lines 6 inches (152 mm) and smaller. Type F steel pipe shall not be used for refrigerant lines having an operating temperature less than -20°F (-29°C).

    1107.4.2 Copper and brass pipe. Standard iron-pipe size, copper and red brass (not less than 80-percent copper) pipe shall conform to ASTM B 42 and ASTM B 43.

    1107.4.3 Copper tube. Copper tube used for refrigerant piping erected on the premises shall be seamless copper tube of Type ACR (hard or annealed) complying with ASTM B 280. Where approved, copper tube for refrigerant piping erected on the premises shall be seamless copper tube of Type K, L or M (drawn or annealed) in accordance with ASTM B 88. Annealed temper copper tube shall not be used in sizes larger than a 2-inch (51 mm) nominal size. Mechanical joints shall not be used on annealed temper copper tube in sizes larger than $^7/_8$-inch (22.2 mm) OD size.

    1107.4.4 Copper tubing joints. Copper tubing joints used in refrigerating systems containing Group A2, A3, B2 or B3 refrigerants shall be brazed. Soldered joints shall not be used in such refrigerating systems.

    1107.4.5 Aluminum tube. Type 3003-0 aluminum tubing with high-pressure fittings shall not be used with methyl chloride and other refrigerants known to attack aluminum.

1107.5 Joints and refrigerant-containing parts in air ducts. Joints and all refrigerant-containing parts of a refrigerating system located in an air duct of an air-conditioning system carrying conditioned air to and from human-occupied space shall be constructed to withstand, without leakage, a pressure of 150 percent of the higher of the design pressure or pressure relief device setting.

1107.6 Exposure of refrigerant pipe joints. Refrigerant pipe joints erected on the premises shall be exposed for visual inspection prior to being covered or enclosed.

1107.7 Stop valves. All systems containing more than 6.6 pounds (3 kg) of a refrigerant in systems using positive-displacement compressors shall have stop valves installed as follows:

1. At the inlet of each compressor, compressor unit or condensing unit.

2. At the discharge outlet of each compressor, compressor unit or condensing unit and of each liquid receiver.

Exceptions:

1. Systems that have a refrigerant pumpout function capable of storing the entire refrigerant charge in a receiver or heat exchanger.

2. Systems that are equipped with provisions for pumpout of the refrigerant using either portable or permanently installed recovery equipment.

3. Self-contained systems.

1107.7.1 Liquid receivers. All systems containing 100 pounds (45 kg) or more of a refrigerant, other than systems utilizing nonpositive displacement compressors, shall have stop valves, in addition to those required by Section 1107.7, on each inlet of each liquid receiver. Stop valves shall not be required on the inlet of a receiver in a condensing unit, nor on the inlet of a receiver which is an integral part of the condenser.

1107.7.2 Copper tubing. Stop valves used with soft annealed copper tubing or hard-drawn copper tubing $^7/_8$-inch (22.2 mm) OD standard size or smaller shall be securely mounted, independent of tubing fastenings or supports.

1107.7.3 Identification. Stop valves shall be identified where their intended purpose is not obvious. Numbers shall not be used to label the valves, unless a key to the numbers is located near the valves.

SECTION 1108
FIELD TEST

1108.1 General. Every refrigerant-containing part of every system that is erected on the premises, except compressors, condensers, vessels, evaporators, safety devices, pressure gauges and control mechanisms that are listed and factory tested, shall be tested and proved tight after complete installation, and before operation. Tests shall include both the high- and low-pressure sides of each system at not less than the lower of the design pressures or the setting of the pressure relief device(s). The design pressures for testing shall be those listed on the condensing unit, compressor or compressor unit name-plate, as required by ASHRAE 15.

Exceptions:

1. Gas bulk storage tanks that are not permanently connected to a refrigeration system.

2. Systems erected on the premises with copper tubing not exceeding $^5/_8$-inch (15.8 mm) OD, with wall thickness as required by ASHRAE 15, shall be tested in accordance with Section 1108.1, or by means of refrigerant charged into the system at the saturated vapor pressure of the refrigerant at 70°F (21°C) or higher.

3. Limited-charge systems equipped with a pressure relief device, erected on the premises, shall be tested at a pressure not less than one and one-half times the pressure setting of the relief device. If the equipment or appliance has been tested by the manufacturer at one and one-half times the design pressure, the test after erection on the premises shall be conducted at the design pressure.

1108.1.1 Booster compressor. Where a compressor is used as a booster to obtain an intermediate pressure and discharges into the suction side of another compressor shall be considered a part of the low side, provided that it is protected by a pressure relief device.

1108.1.2 Centrifugal/nonpositive displacement compressors. In field-testing systems using centrifugal or other nonpositive displacement compressors, the entire system shall be considered as the low-side pressure for field test purposes.

1108.2 Test gases. Tests shall be performed with an inert dried gas including, but not limited to, nitrogen and carbon dioxide. Oxygen, air, combustible gases and mixtures containing such gases shall not be used.

Exception: The use of air is allowed to test R-717, ammonia, systems provided that they are subsequently evacuated before charging with refrigerant.

1108.3 Test apparatus. The means used to build up the test pressure shall have either a pressure-limiting device or a pressure-reducing device and a gauge on the outlet side.

1108.4 Declaration. A certificate of test shall be provided for all systems containing 55 pounds (25 kg) or more of refrigerant. The certificate shall give the name of the refrigerant and the field test pressure applied to the high side and the low side of the system. The certification of test shall be signed by the installer and shall be made part of the public record.

[F]SECTION 1109
PERIODIC TESTING

1109.1 Testing required. The following emergency devices and systems shall be periodically tested in accordance with the manufacturer's instructions and as required by the code official:

1. Treatment and flaring systems.

2. Valves and appurtenances necessary to the operation of emergency refrigeration control boxes.

3. Fans and associated equipment intended to operate emergency pure ventilation systems.

4. Detection and alarm systems.

CHAPTER 12
HYDRONIC PIPING

SECTION 1201
GENERAL

1201.1 Scope. The provisions of this chapter shall govern the construction, installation, alteration and repair of hydronic piping systems. This chapter shall apply to hydronic piping systems that are part of heating, ventilation and air-conditioning systems. Such piping systems shall include steam, hot water, chilled water, steam condensate and ground source heat pump loop systems. Potable cold and hot water distribution systems shall be installed in accordance with the *International Plumbing Code.*

1201.2 Pipe sizing. Piping for hydronic systems shall be sized for the demand of the system.

SECTION 1202
MATERIAL

1202.1 Piping. Piping material shall conform to the standards cited in this section.

Exception: Embedded piping regulated by Section 1209.

1202.2 Used materials. Reused pipe, fittings, valves or other materials shall be clean and free of foreign materials and shall be approved by the code official for reuse.

1202.3 Material rating. Materials shall be rated for the operating temperature and pressure of the hydronic system. Materials shall be suitable for the type of fluid in the hydronic system.

1202.4 Piping materials standards. Hydronic pipe shall conform to the standards listed in Table 1202.4. The exterior of the pipe shall be protected from corrosion and degradation.

TABLE 1202.4
HYDRONIC PIPE

MATERIAL	STANDARD (see Chapter 15)
Acrylonitrile butadiene styrene (ABS) plastic pipe	ASTM D 1527; ASTM D 2282
Brass pipe	ASTM B 43
Brass tubing	ASTM B 135
Copper or copper-alloy pipe	ASTM B 42; ASTM B 302
Copper or copper-alloy tube (Type K, L or M)	ASTM B 75; ASTM B 88; ASTM B 251
Chlorinated polyvinyl chloride (CPVC) plastic pipe	ASTM D 2846; ASTM F 441; ASTM F 442
Cross-linked polyethylene/ aluminum/cross-linked polyethylene (PEX-AL-PEX) pressure pipe	ASTM F 1281; CSA CAN/CSA-B-137.10

(continued)

TABLE 1202.4—continued
HYDRONIC PIPE

MATERIAL	STANDARD (see Chapter 15)
Cross-linked polyethylene (PEX) tubing	ASTM F 876; ASTM F 877
Lead pipe	FS WW-P-325B
Polybutylene (PB) plastic pipe and tubing	ASTM D 3309
Polyethylene (PE) pipe, tubing and fittings (for ground source heat pump loop systems)	ASTM D 2513; ASTM D 3035; ASTM D 2447; ASTM D 2683; ASTM F 1055; ASTM D 2837; ASTM D 3350; ASTM D 1693
Polyvinyl chloride (PVC) plastic pipe	ASTM D 1785; ASTM D 2241
Steel pipe	ASTM A 53; ASTM A 106
Steel tubing	ASTM A 254

1202.5 Pipe fittings. Hydronic pipe fittings shall be approved for installation with the piping materials to be installed, and shall conform to the respective pipe standards or to the standards listed in Table 1202.5.

TABLE 1202.5
HYDRONIC PIPE FITTINGS

MATERIAL	STANDARD (see Chapter 15)
Bronze	ASME B16.24
Copper and copper alloys	ASME B16.15; ASME B16.18; ASME B16.22; ASME B16.23; ASME B16.26; ASME B16.29
Gray iron	ASTM A 126
Malleable iron	ASME B16.3
Plastic	ASTM D 2466; ASTM D 2467; ASTM D 2468; ASTM F 438; ASTM F 439; ASTM F 877
Steel	ASME B16.5; ASME B16.9; ASME B16.11; ASME B16.28; ASTM A 420
Brass	ASTM F 1974

1202.6 Valves. Valves shall be constructed of materials that are compatible with the type of piping material and fluids in the system. Valves shall be rated for the temperatures and pressures of the systems in which the valves are installed.

1202.7 Flexible connectors, expansion and vibration compensators. Flexible connectors, expansion and vibration control devices and fittings shall be of an approved type.

SECTION 1203
JOINTS AND CONNECTIONS

1203.1 Approval. Joints and connections shall be of an approved type. Joints and connections shall be tight for the pressure of the hydronic system.

1203.1.1 Joints between different piping materials. Joints between different piping materials shall be made with approved adapter fittings. Joints between different metallic piping materials shall be made with approved dielectric fittings or brass converter fittings.

1203.2 Preparation of pipe ends. Pipe shall be cut square, reamed and chamfered, and shall be free of burrs and obstructions. Pipe ends shall have full-bore openings and shall not be undercut.

1203.3 Joint preparation and installation. When required by Sections 1203.4 through 1203.14, the preparation and installation of brazed, mechanical, soldered, solvent-cemented, threaded and welded joints shall comply with Sections 1203.3.1 through 1203.3.7.

1203.3.1 Brazed joints. Joint surfaces shall be cleaned. An approved flux shall be applied where required. The joint shall be brazed with a filler metal conforming to AWS A5.8.

1203.3.2 Mechanical joints. Mechanical joints shall be installed in accordance with the manufacturer's instructions.

1203.3.3 Soldered joints. Joint surfaces shall be cleaned. A flux conforming to ASTM B 813 shall be applied. The joint shall be soldered with a solder conforming to ASTM B 32.

1203.3.4 Solvent-cemented joints. Joint surfaces shall be clean and free of moisture. An approved primer shall be applied to CPVC and PVC pipe-joint surfaces. Joints shall be made while the cement is wet. Solvent cement conforming to the following standards shall be applied to all joint surfaces:

1. ASTM D 2235 for ABS joints.

2. ASTM F 493 for CPVC joints.

3. ASTM D 2564 for PVC joints.

   CPVC joints shall be made in accordance with ASTM D 2846.

1203.3.5 Threaded joints. Threads shall conform to ASME B1.20.1. Schedule 80 or heavier plastic pipe shall be threaded with dies specifically designed for plastic pipe. Thread lubricant, pipe-joint compound or tape shall be applied on the male threads only and shall be approved for application on the piping material.

1203.3.6 Welded joints. Joint surfaces shall be cleaned by an approved procedure. Joints shall be welded with an approved filler metal.

1203.3.7 Grooved and shouldered joints. Grooved and shouldered joints shall be approved and installed in accordance with the manufacturer's installation instructions.

1203.3.8 Mechanically formed tee fittings. Mechanically extracted outlets shall have a height not less than three times the thickness of the branch tube wall.

1203.3.8.1 Full flow assurance. Branch tubes shall not restrict the flow in the run tube. A dimple/depth stop shall be formed in the branch tube to ensure that penetration into the outlet is of the correct depth. For inspection purposes, a second dimple shall be placed 0.25 inch (6.4 mm) above the first dimple. Dimples shall be aligned with the tube run.

1203.3.8.2 Brazed joints. Mechanically formed tee fittings shall be brazed in accordance with Section 1203.3.1.

1203.4 ABS plastic pipe. Joints between ABS plastic pipe or fittings shall be solvent-cemented or threaded joints conforming to Section 1203.3.

1203.5 Brass pipe. Joints between brass pipe or fittings shall be brazed, mechanical, threaded or welded joints conforming to Section 1203.3.

1203.6 Brass tubing. Joints between brass tubing or fittings shall be brazed, mechanical or soldered joints conforming to Section 1203.3.

1203.7 Copper or copper-alloy pipe. Joints between copper or copper-alloy pipe or fittings shall be brazed, mechanical, soldered, threaded or welded joints conforming to Section 1203.3.

1203.8 Copper or copper-alloy tubing. Joints between copper or copper-alloy tubing or fittings shall be brazed, mechanical or soldered joints conforming to Section 1203.3 or flared joints conforming to Section 1203.8.1.

1203.8.1 Flared joints. Flared joints shall be made by a tool designed for that operation.

1203.9 CPVC plastic pipe. Joints between CPVC plastic pipe or fittings shall be solvent-cemented or threaded joints conforming to Section 1203.3.

1203.10 Polybutylene plastic pipe and tubing. Joints between polybutylene plastic pipe and tubing or fittings shall be mechanical joints conforming to Section 1203.3 or heat-fusion joints conforming to Section 1203.10.1.

1203.10.1 Heat-fusion joints. Joints shall be of the socket-fusion or butt-fusion type. Joint surfaces shall be clean and free of moisture. Joint surfaces shall be heated to melt temperatures and joined. The joint shall be undisturbed until cool. Joints shall be made in accordance with ASTM D 3309.

1203.11 Cross-linked polyethylene (PEX) plastic tubing. Joints between cross-linked polyethylene plastic tubing and fittings shall conform to Sections 1203.11.1 and 1203.11.2. Mechanical joints shall conform to Section 1203.3.

1203.11.1 Compression-type fittings. When compression-type fittings include inserts and ferrules or O-rings, the fittings shall be installed without omitting the inserts and ferrules or O-rings.

1203.11.2 Plastic-to-metal connections. Soldering on the metal portion of the system shall be performed at least 18 inches (457 mm) from a plastic-to-metal adapter in the same water line.

1203.12 PVC plastic pipe. Joints between PVC plastic pipe and fittings shall be solvent-cemented or threaded joints conforming to Section 1203.3.

1203.13 Steel pipe. Joints between steel pipe or fittings shall be mechanical joints that are made with an approved elastomeric seal, or shall be threaded or welded joints conforming to Section 1203.3.

1203.14 Steel tubing. Joints between steel tubing or fittings shall be mechanical or welded joints conforming to Section 1203.3.

1203.15 Polyethylene plastic pipe and tubing for ground source heat pump loop systems. Joints between polyethylene plastic pipe and tubing or fittings for ground source heat pump loop systems shall be heat fusion joints conforming to Section 1203.15.1, electrofusion joints conforming to Section 1203.15.2, or stab-type insertion joints conforming to Section 1203.15.3.

1203.15.1 Heat-fusion joints. Joints shall be of the socket-fusion, saddle-fusion or butt-fusion type, fabricated in accordance with the piping manufacturer's instructions. Joint surfaces shall be clean and free of moisture. Joint surfaces shall be heated to melt temperatures and joined. The joint shall be undisturbed until cool. Fittings shall be manufactured in accordance with ASTM D 2683.

1203.15.2 Electrofusion joints. Joints shall be of the electrofusion type. Joint surfaces shall be clean and free of moisture, and scoured to expose virgin resin. Joint surfaces shall be heated to melt temperatures for the period of time specified by the manufacturer. The joint shall be undisturbed until cool. Fittings shall be manufactured in accordance with ASTM F 1055.

1203.15.3 Stab-type insert fittings. Joint surfaces shall be clean and free of moisture. Pipe ends shall be chamfered and inserted into the fittings to full depth. Fittings shall be manufactured in accordance with ASTM D 2513.

SECTION 1204
PIPE INSULATION

1204.1 Insulation characteristics. Pipe insulation installed in buildings shall conform to the requirements of the *International Energy Conservation Code*, shall be tested in accordance with ASTM E 84 and shall have a maximum flame spread index of 25 and a smoke-developed index not exceeding 450. Insulation installed in an air plenum shall comply with Section 602.2.1.

Exception: The maximum flame spread index and smoke-developed index shall not apply to one- and two-family dwellings.

1204.2 Required thickness. Hydronic piping shall be insulated to the thickness required by the *International Energy Conservation Code*.

SECTION 1205
VALVES

1205.1 Where required. Shutoff valves shall be installed in hydronic piping systems in the locations indicated in Sections 1205.1.1 through 1205.1.6.

1205.1.1 Heat exchangers. Shutoff valves shall be installed on the supply and return side of a heat exchanger.

Exception: Shutoff valves shall not be required when heat exchangers are integral with a boiler; or are a component of a manufacturer's boiler and heat exchanger packaged unit and are capable of being isolated from the hydronic system by the supply and return valves required by Section 1005.1.

1205.1.2 Central systems. Shutoff valves shall be installed on the building supply and return of a central utility system.

1205.1.3 Pressure vessels. Shutoff valves shall be installed on the connection to any pressure vessel.

1205.1.4 Pressure-reducing valves. Shutoff valves shall be installed on both sides of a pressure-reducing valve.

1205.1.5 Equipment and appliances. Shutoff valves shall be installed on connections to mechanical equipment and appliances. This requirement does not apply to components of a hydronic system such as pumps, air separators, metering devices and similar equipment.

1205.1.6 Expansion tanks. Shutoff valves shall be installed at connections to nondiaphragm-type expansion tanks.

1205.2 Reduced pressure. A pressure relief valve shall be installed on the low-pressure side of a hydronic piping system that has been reduced in pressure. The relief valve shall be set at the maximum pressure of the system design. The valve shall be installed in accordance with Section 1006.

SECTION 1206
PIPING INSTALLATION

1206.1 General. Piping, valves, fittings and connections shall be installed in accordance with the conditions of approval.

1206.1.1 Prohibited tee applications. Fluid in the supply side of a hydronic system shall not enter a tee fitting through the branch opening.

1206.2 System drain down. Hydronic piping systems shall be designed and installed to permit the system to be drained. Where the system drains to the plumbing drainage system, the installation shall conform to the requirements of the *International Plumbing Code*.

1206.3 Protection of potable water. The potable water system shall be protected from backflow in accordance with the *International Plumbing Code*.

1206.4 Pipe penetrations. Openings for pipe penetrations in walls, floors or ceilings shall be larger than the penetrating pipe. Openings through concrete or masonry building elements shall be sleeved. The annular space surrounding pipe penetrations shall be protected in accordance with the *International Building Code*.

1206.5 Clearance to combustibles. A pipe in a hydronic piping system in which the exterior temperature exceeds 250°F (121°C) shall have a minimum clearance of 1 inch (25 mm) to combustible materials.

1206.6 Contact with building material. A hydronic piping system shall not be in direct contact with building materials that cause the piping material to degrade or corrode, or that interfere with the operation of the system.

1206.7 Water hammer. The flow velocity of the hydronic piping system shall be controlled to reduce the possibility of water hammer. Where a quick-closing valve creates water hammer, an approved water-hammer arrestor shall be installed. The arrestor shall be located within a range as specified by the manufacturer of the quick-closing valve.

1206.8 Steam piping pitch. Steam piping shall be installed to drain to the boiler or the steam trap. Steam systems shall not have drip pockets that reduce the capacity of the steam piping.

1206.9 Strains and stresses. Piping shall be installed so as to prevent detrimental strains and stresses in the pipe. Provisions shall be made to protect piping from damage resulting from expansion, contraction and structural settlement. Piping shall be installed so as to avoid structural stresses or strains within building components.

   1206.9.1 Flood hazard. Piping located in a flood-hazard zone or high-hazard zone shall be capable of resisting hydrostatic and hydrodynamic loads and stresses, including the effects of buoyancy, during the occurrence of flooding to the base flood elevation.

1206.10 Pipe support. Pipe shall be supported in accordance with Section 305.

1206.11 Condensation. Provisions shall be made to prevent the formation of condensation on the exterior of piping.

SECTION 1207
TRANSFER FLUID

1207.1 Flash point. The flash point of transfer fluid in a hydronic piping system shall be a minimum of 50°F (28°C) above the maximum system operating temperature.

1207.2 Makeup water. The transfer fluid shall be compatible with the makeup water supplied to the system.

SECTION 1208
TESTS

1208.1 General. Hydronic piping systems other than ground-source heat pump loop systems shall be tested hydrostatically at one and one half times the maximum system design pressure, but not less than 100 psi (689 kPa). The duration of each test shall be not less than 15 minutes. Ground-source heat pump loop systems shall be tested in accordance with Section 1208.1.1.

   1208.1.1 Ground source heat pump loop systems. Before connection (header) trenches are backfilled, the assembled loop system shall be pressure tested with water at 100 psi (689 kPa) for 30 minutes with no observed leaks. Flow and pressure loss testing shall be performed and the actual flow rates and pressure drops shall be compared to the calculated design values. If actual flow rate or pressure drop values differ from calculated design values by more than 10 percent, the problem shall be identified and corrected.

SECTION 1209
EMBEDDED PIPING

1209.1 Materials. Piping for heating panels shall be standard-weight steel pipe, Type L copper tubing, polybutylene or other approved plastic pipe or tubing rated at 100 psi (689 kPa) at 180°F (82°C).

1209.2 Pressurizing during installation. Piping to be embedded in concrete shall be pressure tested prior to pouring concrete. During pouring, the pipe shall be maintained at the proposed operating pressure.

1209.3 Embedded joints. Joints of pipe or tubing that are embedded in a portion of the building, such as concrete or plaster, shall be in accordance with the requirements of Sections 1209.3.1 through 1209.3.3.

   1209.3.1 Steel pipe joints. Steel pipe shall be welded by electrical arc or oxygen/acetylene method.

   1209.3.2 Copper tubing joints. Copper tubing shall be joined by brazing with filler metals having a melting point of not less than 1,000°F (538°C).

   1209.3.3 Polybutylene joints. Polybutylene pipe and tubing shall be installed in continuous lengths or shall be joined by heat fusion in accordance with Section 1203.10.1.

1209.4 Not embedded related piping. Joints of other piping in cavities or running exposed shall be joined by approved methods in accordance with manufacturer's installation instructions and related sections of this code.

CHAPTER 13

FUEL OIL PIPING AND STORAGE

SECTION 1301
GENERAL

1301.1 Scope. This chapter shall govern the design, installation, construction and repair of fuel-oil storage and piping systems. The storage of fuel oil and flammable and combustible liquids shall be in accordance with the *International Fire Code*.

1301.2 Storage and piping systems. Fuel-oil storage systems shall comply with the *International Fire Code*. Fuel-oil piping systems shall comply with the requirements of this code.

1301.3 Fuel type. An appliance shall be designed for use with the type of fuel to which it will be connected. Such appliance shall not be converted from the fuel specified on the rating plate for use with a different fuel without securing reapproval from the code official.

1301.4 Fuel tanks, piping and valves. The tank, piping and valves for appliances burning oil shall be installed in accordance with the requirements of this chapter. When an oil burner is served by a tank, any part of which is above the level of the burner inlet connection and where the fuel supply line is taken from the top of the tank, an approved antisiphon valve or other siphon-breaking device shall be installed in lieu of the shutoff valve.

SECTION 1302
MATERIAL

1302.1 General. Piping materials shall conform to the standards cited in this section.

1302.2 Rated for system. All materials shall be rated for the operating temperatures and pressures of the system, and shall be compatible with the type of liquid.

1302.3 Pipe standards. Fuel oil pipe shall comply with one of the standards listed in Table 1302.3.

TABLE 1302.3
FUEL OIL PIPING

MATERIAL	STANDARD (see Chapter 15)
Brass pipe	ASTM B 43
Brass tubing	ASTM B 135
Copper or copper-alloy pipe	ASTM B 42; ASTM B 302
Copper or copper-alloy tubing (Type K, L or M)	ASTM B 75; ASTM B 88; ASTM B 280
Labeled pipe	(See Section 1302.4)
Nonmetallic pipe	ASTM D 2996
Steel pipe	ASTM A 53; ASTM A 106
Steel tubing	ASTM A 254; ASTM A 539

1302.4 Nonmetallic pipe. All nonmetallic pipe shall be listed and labeled as being acceptable for the intended application for flammable and combustible liquids. Nonmetallic pipe shall be installed only outside, underground.

1302.5 Fittings and valves. Fittings and valves shall be approved for the piping systems, and shall be compatible with, or shall be of the same material as, the pipe or tubing.

1302.6 Bending of pipe. Pipe shall be approved for bending. Pipe bends shall be made with approved equipment. The bend shall not exceed the structural limitations of the pipe.

1302.7 Pumps. Pumps that are not part of an appliance shall be of a positive-displacement type. The pump shall automatically shut off the supply when not in operation. Pumps shall be listed and labeled in accordance with UL 343.

1302.8 Flexible connectors and hoses. Flexible connectors and hoses shall be listed and labeled in accordance with UL 536.

SECTION 1303
JOINTS AND CONNECTIONS

1303.1 Approval. Joints and connections shall be approved and of a type approved for fuel-oil piping systems. All threaded joints and connections shall be made tight with suitable lubricant or pipe compound. Unions requiring gaskets or packings, right or left couplings, and sweat fittings employing solder having a melting point of less than 1,000°F (538°C) shall not be used in oil lines. Cast-iron fittings shall not be used. Joints and connections shall be tight for the pressure required by test.

1303.1.1 Joints between different piping materials. Joints between different piping materials shall be made with approved adapter fittings. Joints between different metallic piping materials shall be made with approved dielectric fittings or brass converter fittings.

1303.2 Preparation of pipe ends. All pipe shall be cut square, reamed and chamfered and be free of all burrs and obstructions. Pipe ends shall have full-bore openings and shall not be undercut.

1303.3 Joint preparation and installation. Where required by Sections 1303.4 through 1303.10, the preparation and installation of brazed, mechanical, threaded and welded joints shall comply with Sections 1303.3.1 through 1303.3.4.

1303.3.1 Brazed joints. All joint surfaces shall be cleaned. An approved flux shall be applied where required. The joints shall be brazed with a filler metal conforming to AWS A5.8.

1303.3.2 Mechanical joints. Mechanical joints shall be installed in accordance with the manufacturer's instructions.

1303.3.3 Threaded joints. Threads shall conform to ASME B1.20.1. Pipe-joint compound or tape shall be applied on the male threads only.

1303.3.4 Welded joints. All joint surfaces shall be cleaned by approved procedure. The joint shall be welded with an approved filler metal.

1303.4 Brass pipe. Joints between brass pipe or fittings shall be brazed, mechanical, threaded or welded joints complying with Section 1303.3.

1303.5 Brass tubing. Joints between brass tubing or fittings shall be brazed or mechanical joints complying with Section 1303.3.

1303.6 Copper or copper-alloy pipe. Joints between copper or copper-alloy pipe or fittings shall be brazed, mechanical, threaded or welded joints complying with Section 1303.3.

1303.7 Copper or copper-alloy tubing. Joints between copper or copper-alloy tubing or fittings shall be brazed or mechanical joints complying with Section 1303.3 or flared joints. Flared joints shall be made by a tool designed for that operation.

1303.8 Nonmetallic pipe. Joints between nonmetallic pipe or fittings shall be installed in accordance with the manufacturer's instructions for the labeled pipe and fittings.

1303.9 Steel pipe. Joints between steel pipe or fittings shall be threaded or welded joints complying with Section 1303.3 or mechanical joints complying with Section 1303.9.1.

1303.9.1 Mechanical joints. Joints shall be made with an approved elastomeric seal. Mechanical joints shall be installed in accordance with the manufacturer's instructions. Mechanical joints shall be installed outside, underground, unless otherwise approved.

1303.10 Steel tubing. Joints between steel tubing or fittings shall be mechanical or welded joints complying with Section 1303.3.

1303.11 Piping protection. Proper allowance shall be made for expansion, contraction, jarring and vibration. Piping other than tubing, connected to underground tanks, except straight fill lines and test wells, shall be provided with flexible connectors, or otherwise arranged to permit the tanks to settle without impairing the tightness of the piping connections.

SECTION 1304
PIPING SUPPORT

1304.1 General. Pipe supports shall be in accordance with Section 305.

SECTION 1305
FUEL OIL SYSTEM INSTALLATION

1305.1 Size. The fuel oil system shall be sized for the maximum capacity of fuel oil required. The minimum size of a supply line shall be $^3/_8$-inch (9.5 mm) inside diameter nominal pipe or $^3/_8$-inch (9.5 mm) OD tubing. The minimum size of a return line shall be $^1/_4$-inch (6.4 mm) inside diameter nominal pipe or $^5/_{16}$-inch (7.9 mm) outside diameter tubing. Copper tubing shall

have 0.035-inch (0.9 mm) nominal and 0.032-inch (0.8 mm) minimum wall thickness.

1305.2 Protection of pipe, equipment and appliances. All fuel oil pipe, equipment and appliances shall be protected from physical damage.

1305.2.1 Flood hazard. All fuel oil pipe, equipment and appliances located in flood hazard areas shall be located above the design flood elevation or shall be capable of resisting hydrostatic and hydrodynamic loads and stresses, including the effects of buoyancy, during the occurrence of flooding to the design flood elevation.

1305.3 Supply piping. Supply piping shall connect to the top of the fuel oil tank. Fuel oil shall be supplied by a transfer pump or automatic pump or by other approved means.

Exception: This section shall not apply to inside or above-ground fuel oil tanks.

1305.4 Return piping. Return piping shall connect to the top of the fuel oil tank. Valves shall not be installed on return piping.

1305.5 System pressure. The system shall be designed for the maximum pressure required by the fuel-oil-burning appliance. Air or other gases shall not be used to pressurize tanks.

1305.6 Fill piping. A fill pipe shall terminate outside of a building at a point at least 2 feet (610 mm) from any building opening at the same or lower level. A fill pipe shall terminate in a manner designed to minimize spilling when the filling hose is disconnected. Fill opening shall be equipped with a tight metal cover designed to discourage tampering.

1305.7 Vent piping. Liquid fuel vent pipes shall terminate outside of buildings at a point not less than 2 feet (610 mm) measured vertically or horizontally from any building opening. Outer ends of vent pipes shall terminate in a weatherproof vent cap or fitting or be provided with a weatherproof hood. All vent caps shall have a minimum free open area equal to the cross-sectional area of the vent pipe and shall not employ screens finer than No. 4 mesh. Vent pipes shall terminate sufficiently above the ground to avoid being obstructed with snow or ice. Vent pipes from tanks containing heaters shall be extended to a location where oil vapors discharging from the vent will be readily diffused. If the static head with a vent pipe filled with oil exceeds 10 pounds per square inch (psi) (69 kPa), the tank shall be designed for the maximum static head that will be imposed.

Liquid fuel vent pipes shall not be cross connected with fill pipes, lines from burners or overflow lines from auxiliary tanks.

SECTION 1306
OIL GAUGING

1306.1 Level indication. All tanks in which a constant oil level is not maintained by an automatic pump shall be equipped with a method of determining the oil level.

1306.2 Test wells. Test wells shall not be installed inside buildings. For outside service, test wells shall be equipped with a tight metal cover designed to discourage tampering.

1306.3 Inside tanks. The gauging of inside tanks by means of measuring sticks shall not be permitted. An inside tank provided with fill and vent pipes shall be provided with a device to indicate either visually or audibly at the fill point when the oil in the tank has reached a predetermined safe level.

1306.4 Gauging devices. Gauging devices such as liquid level indicators or signals shall be designed and installed so that oil vapor will not be discharged into a building from the liquid fuel supply system.

1306.5 Gauge glass. A tank used in connection with any oil burner shall not be equipped with a glass gauge or any gauge which, when broken, will permit the escape of oil from the tank.

SECTION 1307
FUEL OIL VALVES

1307.1 Building shutoff. A shutoff valve shall be installed on the fuel-oil supply line at the entrance to the building. Inside or above-ground tanks are permitted to have valves installed at the tank. The valve shall be capable of stopping the flow of fuel oil to the building or to the appliance served where the valve is installed at a tank inside the building.

1307.2 Appliance shutoff. A shutoff valve shall be installed at the connection to each appliance where more than one fuel-oil-burning appliance is installed.

1307.3 Pump relief valve. A relief valve shall be installed on the pump discharge line where a valve is located downstream of the pump and the pump is capable of exceeding the pressure limitations of the fuel oil system.

1307.4 Fuel-oil heater relief valve. A relief valve shall be installed on the discharge line of fuel-oil-heating appliances.

1307.5 Relief valve operation. The relief valve shall discharge fuel oil when the pressure exceeds the limitations of the system. The discharge line shall connect to the fuel oil tank.

SECTION 1308
TESTING

1308.1 Testing required. Fuel oil piping shall be tested in accordance with NFPA 31.

CHAPTER 14

SOLAR SYSTEMS

SECTION 1401
GENERAL

1401.1 Scope. This chapter shall govern the design, construction, installation, alteration and repair of systems, equipment and appliances intended to utilize solar energy for space heating or cooling, domestic hot water heating, swimming pool heating or process heating.

1401.2 Potable water supply. Potable water supplies to solar systems shall be protected against contamination in accordance with the *International Plumbing Code*.

> **Exception:** Where all solar system piping is a part of the potable water distribution system, in accordance with the requirements of the *International Plumbing Code*, and all components of the piping system are listed for potable water use, cross-connection protection measures shall not be required.

1401.3 Heat exchangers. Heat exchangers used in domestic water-heating systems shall be approved for the intended use. The system shall have adequate protection to ensure that the potability of the water supply and distribution system is properly safeguarded.

1401.4 Solar energy equipment and appliances. Solar energy equipment and appliances shall conform to the requirements of this chapter and shall be installed in accordance with the manufacturer's installation instructions.

1401.5 Ducts. Ducts utilized in solar heating and cooling systems shall be constructed and installed in accordance with Chapter 6 of this code.

SECTION 1402
INSTALLATION

1402.1 Access. Access shall be provided to solar energy equipment and appliances for maintenance. Solar systems and appurtenances shall not obstruct or interfere with the operation of any doors, windows or other building components requiring operation or access.

1402.2 Protection of equipment. Solar equipment exposed to vehicular traffic shall be installed not less than 6 feet (1829 mm) above the finished floor.

> **Exception:** This section shall not apply where the equipment is protected from motor vehicle impact.

1402.3 Controlling condensation. Where attics or structural spaces are part of a passive solar system, ventilation of such spaces, as required by Section 406, is not required where other approved means of controlling condensation are provided.

1402.4 Roof-mounted collectors. Roof-mounted solar collectors that also serve as a roof covering shall conform to the requirements for roof coverings in accordance with the *International Building Code*.

> **Exception:** The use of plastic solar collector covers shall be limited to those approved plastics meeting the requirements for plastic roof panels in the *International Building Code*.

1402.4.1 Collectors mounted above the roof. When mounted on or above the roof covering, the collector array and supporting construction shall be constructed of noncombustible materials or fire-retardant-treated wood conforming to the *International Building Code* to the extent required for the type of roof construction of the building to which the collectors are accessory.

> **Exception:** The use of plastic solar collector covers shall be limited to those approved plastics meeting the requirements for plastic roof panels in the *International Building Code*.

1402.5 Equipment. The solar energy system shall be equipped in accordance with the requirements of Sections 1402.5.1 through 1402.5.4.

1402.5.1 Pressure and temperature. Solar energy system components containing pressurized fluids shall be protected against pressures and temperatures exceeding design limitations with a pressure and temperature relief valve. Each section of the system in which excessive pressures are capable of developing shall have a relief device located so that a section cannot be valved off or otherwise isolated from a relief device. Relief valves shall comply with the requirements of Section 1006.4 and discharge in accordance with Section 1006.6.

1402.5.2 Vacuum. The solar energy system components that are subjected to a vacuum while in operation or during shutdown shall be designed to withstand such vacuum or shall be protected with vacuum relief valves.

1402.5.3 Protection from freezing. System components shall be protected from damage by freezing of heat transfer liquids at the lowest ambient temperatures that will be encountered during the operation of the system.

1402.5.4 Expansion tanks. Liquid single-phase solar energy systems shall be equipped with expansion tanks sized in accordance with Section 1009.

1402.6 Penetrations. Roof and wall penetrations shall be flashed and sealed to prevent entry of water, rodents and insects.

1402.7 Filtering. Air transported to occupied spaces through rock or dust-producing materials by means other than natural convection shall be filtered at the outlet from the heat storage system.

SECTION 1403
HEAT TRANSFER FLUIDS

1403.1 Flash point. The flash point of the actual heat transfer fluid utilized in a solar system shall be not less than 50°F (28°C) above the design maximum nonoperating (no-flow) temperature of the fluid attained in the collector.

1403.2 Flammable gases and liquids. A flammable liquid or gas shall not be utilized as a heat transfer fluid. The flash point of liquids used in occupancies classified in Group H or F shall not be lower unless approved.

SECTION 1404
MATERIALS

1404.1 Collectors. Factory-built collectors shall be listed and labeled, and bear a label showing the manufacturer's name and address, model number, collector dry weight, collector maximum allowable operating and nonoperating temperatures and pressures, minimum allowable temperatures and the types of heat transfer fluids that are compatible with the collector. The label shall clarify that these specifications apply only to the collector.

1404.2 Thermal storage units. Pressurized thermal storage units shall be listed and labeled, and bear a label showing the manufacturer's name and address, model number, serial number, storage unit maximum and minimum allowable operating temperatures, storage unit maximum and minimum allowable operating pressures and the types of heat transfer fluids compatible with the storage unit. The label shall clarify that these specifications apply only to the thermal storage unit.

CHAPTER 15

REFERENCED STANDARDS

This chapter lists the standards that are referenced in various sections of this document. The standards are listed herein by the promulgating agency of the standard, the standard identification, the effective date and title, and the section or sections of this document that reference the standard. The application of the referenced standards shall be as specified in Section 102.8.

ACCA

Air Conditioning Contractors of America
1712 New Hampshire Ave, NW
Washington, DC 20009

Standard Reference Number	Title	Referenced in code section number
Manual D—95	Residential Duct Systems	603.2

ANSI

American National Standards Institute
11 West 42nd Street
New York, NY 10036

Standard reference number	Title	Referenced in code section number
Z21.8—1994	Installation of Domestic Gas Conversion Burners	919.1
Z21.83—1998	Fuel Cell Power Plants	924.1

ARI

Air-Conditioning and Refrigeration Institute
Suite 425
4301 North Fairfax Drive
Arlington, VA 22203

Standard Reference Number	Title	Referenced in code section number
700—95	Specifications for Fluorocarbon and Other Refrigerants	1102.2.2.3

ASHRAE

American Society of Heating, Refrigerating
and Air-Conditioning Engineers, Inc.
1791 Tullie Circle, NE
Atlanta, GA 30329-2305

Standard Reference Number	Title	Referenced in code section number
ASHRAE—2001	ASHRAE Fundamentals Handbook—2001	312.1, 603.2
15—2001	Safety Standard for Refrigeration Systems	1101.6, 1105.8, 1108.1
34—2001	Designation and Safety Classification of Refrigerants	202, 1102.2.1, 1103.1
ASHRAE—2000	HVAC Systems and Equipment Handbook—2000	312.1

ASME

American Society of Mechanical Engineers
Three Park Avenue
New York, NY 10016-5990

Standard Reference Number	Title	Referenced in code section number
B1.20.1—1983 (R1999)	Pipe Threads, General Purpose (Inch)	1203.3.5, 1303.3.3
B16.3—1999	Malleable Iron Threaded Fittings, Classes 150 & 300	Table 1202.5
B16.5—1996	Pipe Flanges and Flanged Fittings NPS ½ through NPS 24— With B16.5a-1998 Addenda	Table 1202.5

ASME—continued

B16.9—1993	Factory Made Wrought Steel Buttwelding Fittings	Table 1202.5
B16.11—1996	Forged Fittings, Socket-Welding and Threaded	Table 1202.5
B16.15—1985(R1994)	Cast Bronze Threaded Fittings	Table 1202.5
B16.18—1984(R1994)	Cast Copper Alloy Solder Joint Pressure Fittings	513.13.1, Table 1202.5
B16.22—1995	Wrought Copper and Copper Alloy Solder Joint Pressure Fittings—with B16.22a-1998 Addenda	513.13.1, Tabl e 1202.5
B16.23—1992	Cast Copper Alloy Solder Joint Drainage Fittings DWV	Table 1202.5
B16.24—1991 (R1998)	Cast Copper Alloy Pipe Flanges and Flanged Fittings: Class 150, 300, 400, 600, 900, 1500 and 2500	Table 1202.5
B16.26—1988	Cast Copper Alloy Fittings for Flared Copper Tubes	Table 1202.5
B16.28—1994	Wrought Steel Buttwelding Short Radius Elbows and Returns	Table 1202.5
B16.29—1994	Wrought Copper and Wrought Copper Alloy Solder Joint Drainage Fittings-DWV	Table 1202.5
BPVC—1998	Boiler & Pressure Vessel Code (Sections I, II, IV, V & VI)	1004.1, 1011.1
CSD-1—1998	Controls and Safety Devices for Automatically Fired Boilers	1004.1

ASTM

ASTM International
100 Barr Harbor Drive
West Conshohocken, PA 19428

Standard Reference Number	Title	Referenced in code section number
A 53/A 53M—01	Specification for Pipe, Steel, Black and Hot-Dipped, Zinc-Coated Welded and Seamless	Table 1202.4, Table 1302.3
A 106—99el	Specification for Seamless Carbon Steel Pipe for High-Temperature Service	Table 1202.4, Table 1302.3
A 126—01	Specification for Gray Iron Castings for Valves, Flanges, and Pipe Fittings	Table 1202.5
A 254—97	Specification for Copper Brazed Steel Tubing	Table 1202.4, Table 1302.3
A 420/A 420M—01	Specification for Piping Fittings of Wrought Carbon Steel and Alloy Steel for Low-Temperature Service	Table 1202.5
A 539—99	Specification for Electric-Resistance-Welded Coiled Steel Tubing for Gas and Fuel Oil Lines	Table 1302.3
B 32—00	Specification for Solder Metal	1203.3.3
B 42—98	Specification for Seamless Copper Pipe, Standard Sizes	513.13.1, 1107.4.2, Table 1202.4, Table 1302.3
B 43—98	Specification for Seamless Red Brass Pipe, Standard Sizes	513.13.1, 1107.4.2, Table 1202.4, Table 1302.3
B 68—99	Specification for Seamless Copper Tube, Bright Annealed	513.13.1
B 75—99	Specification for Seamless Copper Tube	Table 1202.4, Table 1302.3
B 88—99el	Specification for Seamless Copper Water Tube	513.13.1, 1107.4.3, Table 1202.4, Table 1302.3
B 135—00	Specification for Seamless Brass Tube	Table 1202.4, Table 1302.3
B 251—97	Specification for General Requirements for Wrought Seamless Copper and Copper-Alloy Tube	513.13.1, Table 1202.4
B 280—99el	Specification for Seamless Copper Tube for Air Conditioning and Refrigeration Field Service	513.13.1, 1107.4.3, Table 1302.3
B 302—00	Specification for Threadless Copper Pipe, Standard Sizes	Table 1202.4, Table 1302.3
B 813—00e01	Specification for Liquid and Paste Fluxes for Soldering of Copper and Copper Alloy Tube	1203.3.3
C 315—00	Specification for Clay Flue Linings	801.16.1, Table 803.10.4
C 411—97	Test Method for Hot-Surface Performance of High-Temperature Thermal Insulation	604.3
D 56—01	Test Method for Flash Point by Tag Closed Tester	202
D 93—00	Test Method for Flash Point of Pensky-Martens Closed Cup Tester	202
D 1527—99	Specification for Acrylonitrile-Butadiene-Styrene (ABS) Plastic Pipe, Schedules 40 and 80	Table 1202.4
D 1693—01	Test Method for Environmental Stress-Cracking of Ethylene Plastics	Table 1202.4
D 1785—99	Specification for Poly (Vinyl Chloride)(PVC) Plastic Pipe, Schedules 40, 80 and 120	Table 1202.4
D 2235—01	Specifications for Solvent Cement for Acrylonitrile-Butadiene-Styrene (ABS) Plastic Pipe and Fittings	1203.3.4
D 2241—01	Specification for Poly (Vinyl Chloride)(PVC) Pressure-Rated Pipe (SDR-Series)	Table 1202.4
D 2282—99	Specification for Acrylonitrile-Butadiene-Styrene (ABS) Plastic Pipe (SDR-PR)	Table 1202.4
D 2412—96a	Test Method for Determination of External Loading Characteristics of Plastic Pipe by Parallel-Plate Loading	603.8.3
D 2447—99	Specification for Polyethylene (PE) Plastic Pipe, Schedules 40 and 80, Based on Outside Diameter	Table 1202.4
D 2466—01	Specification for Poly (Vinyl Chloride)(PVC) Plastic Pipe Fittings, Schedule 40	Table 1202.5

ASTM—continued

D 2467—01	Specification for Poly (Vinyl Chloride)(PVC) Plastic Pipe Fittings, Schedule 80	Table 1202.5
D 2468—96a	Specification for Acrylonitrile-Butadiene-Styrene (ABS) Plastic Pipe Fittings, Schedule 40	Table 1202.5
D 2513—00	Specification for Thermoplastic Gas Pressure Pipe, Tubing, and Fittings	Table 1202.4, 1203.15.3
D 2564—96a	Specification for Solvent Cements for Poly (Vinyl Chloride) (PVC) Plastic Piping Systems	1203.3.4
D 2683—98	Specification for Socket-Type Polyethylene Fittings for Outside Diameter-Controlled Polyethylene Pipe and Tubing	Table 1202.4, 1203.15.1
D 2837—98a	Test Method for Obtaining Hydrostatic Design Basis for Thermoplastic Pipe Materials	Table 1202.4
D 2846/D 2846M—99	Specification for Chlorinated Poly (Vinyl Chloride) (CPVC) Plastic Hot and Cold Water Distribution Systems	Table 1202.4, 1203.3.4
D 2996—00	Specification for Filament-Wound Fiberglass (Glass Fiber Reinforced Thermosetting Resin) Pipe	Table 1302.3
D 3035—01	Specification for Polyethylene (PE) Plastic Pipe (DR-PR) Based on Controlled Outside Diameter	Table 1202.4
D 3278—96e1	Test Methods for Flash Point of Liquids by Small Scale Closed-Cup Apparatus	202
D 3309—96a	Specification for Polybutylene (PB) Plastic Hot- and Cold- Water Distribution Systems	Table 1202.4, 1203.10.1
D 3350—01	Specification for Polyethylene Plastics Pipe and Fittings Materials	Table 1202.4
E 84—01	Test Method for Surface Burning Characteristics of Building Materials	202, 510.8, 602.2.1, 602.2.1.5, 604.3, 1204.1
E 119—00e	Test Method for Fire Tests of Building Construction and Materials	607.5.2, 607.6.2
E 136—99e01	Test Method for Behavior of Materials in a Vertical Tube Furnace at 750°C	202
E 814—00	Test Method for Fire Tests of Through-Penetration Fire Stops	506.3.10
F 438—01	Specification for Socket Type Chlorinated Poly (Vinyl Chloride) (CPVC) Plastic Pipe Fittings, Schedule 40	Table 1202.5
F 439—01	Specification for Socket Type Chlorinated Poly (Vinyl Chloride) (CPVC) Plastic Pipe Fittings, Schedule 80	Table 1202.5
F 441/F 441M—99	Specification for Chlorinated Poly (Vinyl Chloride) (CPVC) Plastic Pipe, Schedules 40 and 80	Table 1202.4
F 442/F 442M—99	Specification for Chlorinated Poly (Vinyl Chloride) (CPVC) Plastic Pipe (SDR-PR)	Table 1202.4
F 493—97	Specification for Solvent Cements for Chlorinated Poly (Vinyl Chloride) (CPVC) Plastic Pipe and Fittings	1203.3.4
F 876—01	Specification for Crosslinked Polyethylene (PEX) Tubing	Table 1202.4
F 877—01	Specification for Crosslinked Polyethylene (PEX) Plastic Hot and Cold-Water Distribution Systems	Table 1202.4, Table 1202.5
F 1055—98	Specification for Electrofusion Type Polyethylene Fittings for Outside Diameter Controlled Polyethylene Pipe and Tubing	Table 1202.4, 1203.15.2
F 1281—01	Specification for Crosslinked Polyethylene/Aluminum/Crosslinked Polyethylene (PEX-AL-PEX) Pressure Pipe	Table 1202.4
F 1974—00e	Standard Specification for Metal Insert Fittings for Polyethylene/Aluminum/Polyethylene and Crosslinked Polyethylene/Aluminum/Crosslinked Polyethylene Composite Pressure Pipe	Table 1202.5

AWS

American Welding Society
550 N.W. LeJeune Road
P.O. Box 351040
Miami, FL 33135

Standard Reference Number	Title	Referenced in code section number
A5.8—92	Specifications for Filler Metals for Brazing and Braze Welding	1203.3.1, 1303.3.1

CSA

Canadian Standards Association
178 Rexdale Blvd.
Rexdale (Toronto), Ontario, Canada M9W 1R3

Standard Reference Number	Title	Referenced in code section number
CAN/CSA B137.10M—99	Crosslinked Polyethylene/Aluminum/Polyethylene Composite Pressure Pipe Systems	Table 1202.4

DOL

Department of Labor
Occupational Safety and Health Administration
c/o Superintendent of Documents
US Government Printing Office
Washington, DC 20402-9325

Standard Reference Number	Title	Referenced in code section number
29 CFR Part 1910.1000 (1974)	Air Contaminants	502.6

FS

Federal Specifications*
General Services Administration
7th & D Streets
Specification Section, Room 6039
Washington, DC 20407

Standard Reference Number	Title	Referenced in code section number
WW-P-325B (1976)	Pipe, Bends, Traps, Caps and Plugs; Lead (for Industrial Pressure and Soil and Waste Applications	Table 1202.4

* Standards are available from the Supt. of Documents, U.S. Government Printing Office, Washington, DC 20402-9325.

ICC

International Code Council
5203 Leesburg Pike, Suite 708
Falls Church, VA 22041-3401

Standard Reference Number	Title	Referenced in code section number
IBC—03	International Building Code®	201.3, 202, 301.12, 301.12, 301.14, 301.15, 302.1, 302.2, 304.7, 304.10, 308.8, 308.10, 401.4, 401.6, 406.1, 502.10, 502.10.1, 504.2, 506.3.3, 506.3.10, 506.3.12.2, 506.4.1, 509.1, 510.6, 510.6.1, 510.6.2, 510.7, 511.1.5, 513.1, 513.2, 513.3, 513.4.3, 513.5, 513.5.2, 513.5.2.1, 513.6.2, 513.10.5, 513.12, 513.12.2, 513.20, 602.2.1.5.1, 602.2.1.5.2, 602.3, 603.1, 603.10, 604.5.4, 607.1.1, 607.3.2.1, 607.5.1, 607.5.2, 607.5.3, 607.5.4, 607.5.4.1, 607.5.5, 607.5.5.1, 607.6, 701.4.1, 701.4.2, 801.3, 801.16.1, 801.18.4, 902.1, 908.3, 908.4, 910.3, 925.1, 1004.6, 1105.1, 1206.4, 1402.4, 1402.4.1
ICC EC—03	ICC Electrical Code™	201.3, 301.7, 306.3.1, 306.4.1, 513.11, 513.12.1, 602.2.1.1
IEBC—03	International Existing Building Code®	101.2
IECC—03	International Energy Conservation Code®	202, 301.2, 303.3, 312.1, 603.9, 604.1, 1204.1, 1204.2
IFC—03	International Fire Code®	201.3, 310.1, 311.1, 502.5, 502.7.2, 502.8.1, 502.9.5, 502.9.5.2, 502.9.5.3, 502.9.8.2, 502.9.8.3, 502.9.8.5, 502.9.8.6, 502.10, 502.10.3, 502.16.2, 509.1, 510.2.1, 510.2.2, 510.4, 513.12.3, 513.15, 513.16, 513.17, 513.18, 513.19, 513.20.2, 513.20.3, 606.2.1, 908.7, 1101.9, 1105.3, 1106.5, 1106.6, 1301.1, 1301.2
IFGC—03	International Fuel Gas Code®	101.2, 201.3, 301.3, 701.1, 801.1, 901.1, 906.1, 1101.5
IPC—03	International Plumbing Code®	201.3, 301.8, 512.2, 908.5, 1002.1, 1002.2, 1002.3, 1005.2, 1006.6, 1008.2, 1009.3, 1101.4, 1201.1, 1206.2, 1206.3, 1401.2
IRC—03	International Residential Code®	101.2

IIAR

International Institute of Ammonia Refrigeration
Suite 700
1101 Connecticut Ave., NW
Washington, DC 20036

Standard Reference Number	Title	Referenced in code section number
2—99	Equipment, Design, and Installation of Ammonia Mechanical Refrigerating Systems	1101.6

MSS

Manufacturers Standardization Society of the Valve & Fittings Industry, Inc.
127 Park Street, N.E.
Vienna, VA 22180

Standard Reference Number	Title	Referenced in code section number
SP-69—1996	Pipe Hangers and Supports—Selection and Application.	305.4

NAIMA

North American Insulation Manufacturers Association
Suite 310
44 Canal Center Plaza
Alexandria, VA 22314

Standard Reference Number	Title	Referenced in code section number
AH116—02	Fibrous Glass Duct Construction Standards	603.5, 603.9

NFPA

National Fire Protection Association
Batterymarch Park
Quincy, MA 02269

Standard Reference Number	Title	Referenced in code section number
31—01	Installation of Oil-Burning Equipment	801.2.1, 801.18.1, 801.18.2, 920.2, 922.1, 1308.1
37—98	Stationary Combustion Engines and Gas Turbines.	915.1, 915.2
58—01	Liquefied Petroleum Gas Code	502.9.10
69—97	Explosion Prevention Systems	510.8.3
72—99	National Fire Alarm Code	606.3
82—99	Incinerators and Waste and Linen Handling Systems and Equipment	601.1
88B—97	Repair Garages.	304.5
91—99	Exhaust Systems for Air Conveying of Vapors, Gases, Mists, and Noncombustible Particulate Solids	502.9.5.1, 502.17
211—00	Chimneys, Fireplaces, Vents, and Solid Fuel-Burning Appliances	806.1
262—99	Standard Method of Test for Flame Travel and Smoke of Wires and Cables for Use in Air-Handling Spaces	602.2.1.1
704—96	Identification of the Hazards of Materials for Emergency Response	502.8.4, Table 1103.1, 510.1
853—00	Installatin of Stationary Fuel Power Plants	924.1
8501—97	Single Burner Boiler Operation	1004.1
8502—99	Prevention of Furnace Explosions/Implosions in Multiple Burner Boiler-Furnaces.	1004.1
8504—96	Atmospheric Fluidized-Bed Boiler Operation.	1004.1

SMACNA

Sheet Metal & Air Conditioning Contractors National Assoc., Inc.
4021 Fafayette Center Road
Chantilly, VA 22021

Standard Reference Number	Title	Referenced in code section number
SMACNA—95	HVAC Duct Construction Standards—Metal and Flexible	603.4, 603.9
SMACNA—92	Fibrous Glass Duct Construction Standards	603.5, 603.9

UL

Underwriters Laboratories, Inc.
333 Pfingsten Road
Northbrook, IL 60062-2096

Standard Reference Number	Title	Referenced in code section number
17—94	Vent or Chimney Connector Dampers for Oil-Fired Appliances—with Revisions through September 1998	803.6
103—98	Factory-Built Chimneys, Residential Type and Building Heating Appliance—with Revisions through March 1999	805.2
127—96	Factory-Built Fireplaces—with Revisions through November 1999	805.3, 903.1, 903.3
174—98	Household Electric Storage Tank Water Heaters—with revisions through October 1999	1002.1
181—96	Factory-made Air Ducts and Air Connectors—with Revisions through December 1998	512.2, 603.5, 603.6.1, 603.6.2, 604.13
197—93	Commercial Electric Cooking Appliances—With Revisions Through January 2000	507.1
207—93	Refrigerant-Containing Components and Accessories, Nonelectrical—with Revisions Through October 1997	1101.2
343—97	Pumps for Oil-Burning Appliances-with revisions through December 22, 1999	1302.7
391—95	Solid-Fuel and Combination-Fuel Central and Supplementary Furnaces—with Revisions Through May 1999	918.1
412—93	Refrigeration Unit Coolers—with Revisions through November 1998	1101.2
471—95	Commercial Refrigerators and Freezers—with Revisions through April 1998	1101.2
536—97	Flexible metallic Hose - with revisions through October 2000	1302.8
555—99	Fire Dampers - with Revisions through October, 2000	607.3
555C—96	Ceiling Dampers	607.3, 607.6.2
555S—99	Smoke Dampers—with Revisions through December 1999	607.3, 607.3.1.1
586—96	High-Efficiency, Particulate, Air Filter Units - with revisions through april 21, 2000	605.2
641—95	Type L Low-Temperature Venting Systems—with Revisions through April 1999	802.1
710—95	Exhaust Hoods for Commercial Cooking Equipment-with Revisions through April 1999	507.1
726—98	Oil-Fired Boiler Assemblies—with Revisions through January 1999	916.1, 1004.1
727—94	Oil-Fired Centeral Furnaces—with Revisions through January 1999	918.1
729—98	Oil-Fired Floor Furnaces—with Revisions through January 1999	910.1
730—98	Oil-Fired Wall Furnaces—with Revisions through January 1999	909.1
731—95	Oil-Fired Unit Heaters—with Revisions through January 1999	920.1
732—95	Oil-Fired Storage Tank Water Heaters-With revisions through January 1999	1002.1
737—96	Fireplace Stoves—with Revisions through January 2000	805.2, 905.1
762—99	Outline of Investigation for Power Ventilators for Restaurant Exhaust Appliances	506.5.1
791—93	Residential Incinerators—with Revisions through May 1998	907.1
834—98	Heating, Water Supply and Power Boilers Electric—with Revisions Through November 1998	1004.1
867—00	Electrostatic Air Cleaners	605.2
896—93	Oil-Burning Stoves—with Revisions through November 1999	917.1, 922.1
900—94	Air Filter Units - with revisions through October 1999	605.2
959—01	Medium Heat Appliance Factory-Built Chimneys	805.5
1240—94	Electric Commercial Clothes Drying Equipment—with Revisions through October 1999	913.1
1261—96	Electric Water Heaters for Pools and Tubs - with revisions through November 25, 1998	916.1
1453—95	Electronic Booster and Commercial Storage Tank Water Heaters - with Revisions Through September 1998	1002.1
1482—98	Solid-Fuel Type Room Heaters—with Revisions through January 2000	905.1
1777—98	Chimney Liners—with Revisions through July 1998	801.18.4
1820—97	Fire Test of Pneumatic Tubing for Flame and Smoke Characteristics - with Revisions through March 1999	602.2.1.3
1887—96	Fire Tests of Plastic Sprinkler Pipe for Visible Flame and Smoke Characteristics—with Revisions through June 1999	602.2.1.2
1995—98	Heating and Cooling Equipment - with Revisions through August 1999	911.1, 918.1, 918.3, 1101.2
2043—96	Fire Test for Heat and Visible Smoke Release for Discrete Products and their Accessories Installed in Air-Handling Spaces-With Revisions through February 1998	602.2.1.4
2158—97	Outline of Investigation Electric Clothes Dryer—with Revisions through February 1999	913.1
2162—94	Outline of Investigation for Commercial Wood-Fired Baking Ovens—Refractory Type	917.1

APPENDIX A
COMBUSTION AIR OPENINGS AND
CHIMNEY CONNECTOR PASS-THROUGHS

Figures A-1 through A-4 are illustrations of appliances located in confined spaces.

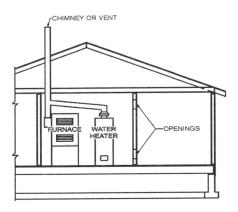

FIGURE A-1
ALL AIR FROM INSIDE THE BUILDING

NOTE: Each opening shall have a free area of not less than 1 square inch per 1,000 Btu per hour of the total input rating of all appliances in the enclosure and not less than 100 square inches.

For SI:1 square inch = 645 mm^2, 1 British thermal unit per hour = 0.2931 W.

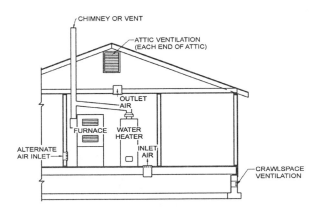

FIGURE A-2
ALL AIR FROM OUTDOORS—INLET AIR FROM VENTILATED CRAWL SPACE AND OUTLET AIR TO VENTILATED ATTIC

NOTE: The inlet and and outlet air openings shall each have a free area of not less than 1 square inch per 4,000 Btu per hour of the total input rating of all appliances in the enclosure.

For SI:1 square inch = 645 mm^2, 1 British thermal unit per hour = 0.2931 W.

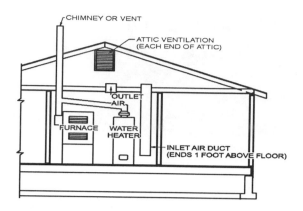

FIGURE A-3
ALL AIR FROM OUTDOORS THROUGH VENTILATED ATTIC

NOTE: The inlet and outlet air openings shall each have a free area of not less than 1 square inch per 4,000 Btu per hour of the total input rating of all appliances in the enclosure.

For SI:1 foot = 304.8 mm, 1 square inch = 645 mm², 1 British thermal unit per hour = 0.2931 W.

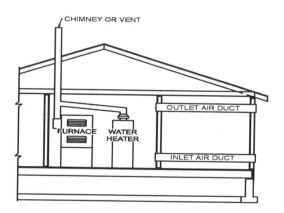

FIGURE A-4
ALL AIR FROM OUTDOORS THROUGH HORIZONTAL DUCTS OR DIRECT OPENINGS

NOTE: Each air duct opening shall have a free area of not less than 1 square inch per 2,000 Btu per hour of the total input rating of all appliances in the enclosure. If the appliance room is located against an outside wall and the air openings communicate directly with the outdoors, each opening shall have a free area of not less than 1 square inch per 4,000 Btu per hour or the total input rating of all appliances in the enclosure.

For SI:1 foot = 304.8 mm, 1 square inch = 645 mm², 1 British thermal unit per hour = 0.2931 W.

SYSTEM A

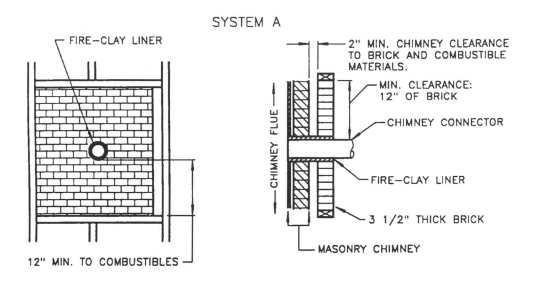

SYSTEM B

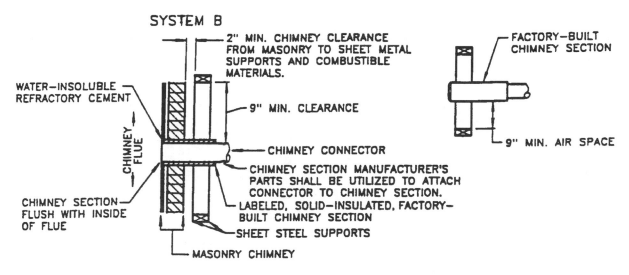

FIGURE A-5
CHIMNEY CONNECTOR SYSTEMS

For SI:1 inch = 25.4 mm.

SYSTEM C

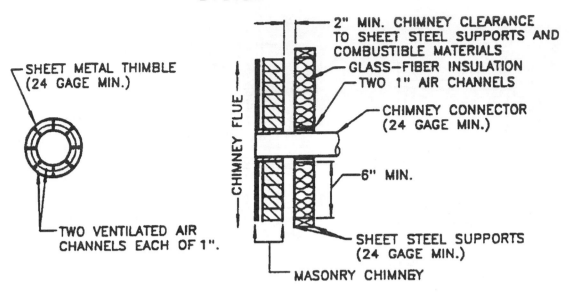

- SHEET METAL THIMBLE (24 GAGE MIN.)
- TWO VENTILATED AIR CHANNELS EACH OF 1".
- 2" MIN. CHIMNEY CLEARANCE TO SHEET STEEL SUPPORTS AND COMBUSTIBLE MATERIALS
- GLASS-FIBER INSULATION
- TWO 1" AIR CHANNELS
- CHIMNEY CONNECTOR (24 GAGE MIN.)
- 6" MIN.
- SHEET STEEL SUPPORTS (24 GAGE MIN.)
- MASONRY CHIMNEY
- CHIMNEY FLUE

SYSTEM D

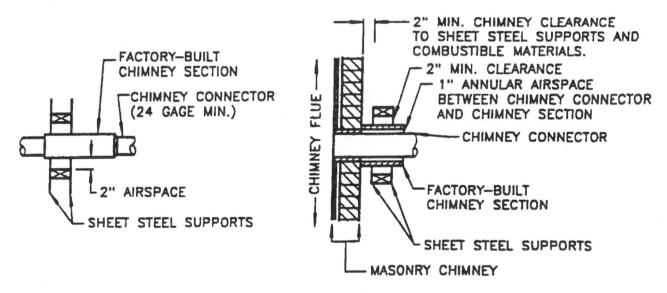

- FACTORY-BUILT CHIMNEY SECTION
- CHIMNEY CONNECTOR (24 GAGE MIN.)
- 2" AIRSPACE
- SHEET STEEL SUPPORTS
- 2" MIN. CHIMNEY CLEARANCE TO SHEET STEEL SUPPORTS AND COMBUSTIBLE MATERIALS.
- 2" MIN. CLEARANCE
- 1" ANNULAR AIRSPACE BETWEEN CHIMNEY CONNECTOR AND CHIMNEY SECTION
- CHIMNEY CONNECTOR
- FACTORY-BUILT CHIMNEY SECTION
- SHEET STEEL SUPPORTS
- MASONRY CHIMNEY
- CHIMNEY FLUE

FIGURE A-5—continued
CHIMNEY CONNECTOR SYSTEMS

For SI:1 inch = 25.4 mm.

APPENDIX B
RECOMMENDED PERMIT FEE SCHEDULE

B101
MECHANICAL WORK, OTHER THAN GAS PIPING SYSTEMS

B101.1 Initial Fee

For issuing each permit $___

B101.2 Additional Fees

B101.2.1 Fee for inspecting heating, ventilating, ductwork, air-conditioning, exhaust, venting, combustion air, pressure vessel, solar, fuel oil and refrigeration systems and appliance installations shall be $___ for the first $1,000.00, or fraction thereof, of valuation of the installation plus $___ for each additional $1,000.00 or fraction thereof.

B101.2.2 Fee for inspecting repairs, alterations and additions to an existing system shall be $___ plus $___ for each $1,000.00 or fraction thereof.

B101.2.3 Fee for inspecting boilers (based upon Btu input):

33,000 Btu (1 BHp) to 165,000 (5 BHp)	$ ___
165,001 Btu (5 BHp) to 330,000 (10 BHp)	$ ___
330,001 Btu (10 BHp) to 1,165,000 (52 BHp)	$ ___
1,165,001 Btu (52 BHp) to 3,300,000 (98 BHp)	$ ___
over 3,300,000 Btu (98 BHp)	$ ___

For SI: 1 British thermal unit = 0.2931 W, 1 BHp = 33,475 Btu/hr.

B102
FEE FOR REINSPECTION

If it becomes necessary to make a reinspection of a heating, ventilation, air-conditioning or refrigeration system, or boiler installation, the installer of such equipment shall pay a reinspection fee of $___.

B103
TEMPORARY OPERATION INSPECTION FEE

When preliminary inspection is requested for purposes of permitting temporary operation of a heating, ventilating, refrigeration, or air-conditioning system, or portion thereof, a fee of $___ shall be paid by the contractor requesting such preliminary inspection. If the system is not approved for temporary operation on the first preliminary inspection, the usual reinspection fee shall be charged for each subsequent preliminary inspection for such purpose.

B104
SELF-CONTAINED UNITS LESS THAN 2 TONS

In all buildings, except one- and two-family dwellings, where self-contained air-conditioning units of less than 2 tons are to be installed, the fee charged shall be that for the total cost of all units combined (see B101.2.1 for rate).

INDEX

EDITORIAL CHANGES – SECOND PRINTING

Page 22, 304.5: line 2 now reads . . . motor fuel-dispensing facilities, repair garages or other areas

Page 28, 403.3.2: Equation 4-1 now reads . . . $Y = \dfrac{X}{(1 + X - Z)}$

Page 39, 506.3.1.2: line 4 now reads . . . with Sections 603.1, 603.3, 603.4, 603.9, 603.10 and 603.12.

Page 59, 607.5.4: line 4 now reads . . . barrier wall or a corridor enclosure required to have smoke